The ROSE HIVE *Method*

Challenging Conventional Beekeeping

Tim Rowe

GREEN HAT BOOKS
www.greenhatbooks.com

The Rose Hive Method
Copyright © 2010 Tim Rowe. All rights reserved. No part of this publication may be reproduced or distributed in any form or by any means without the prior permission of the author/publisher.

ISBN 978-0-9567026-0-9

First Published by Green Hat Books in 2010
2011 reprinted four times.
2012 reprinted four times.
2013 reprinted four times.
This copy printed 2014.

The hive-management practices described in this book are given in good faith but are suggestions only - no responsibility can be taken by the either the author or the publisher for any unforeseen consequences arising from their use.

All photos and illustrations including cover photos copyright © Tim Rowe, except:
Comb (back cover), and Lifting the crownboard (page 4) copyright © Lizzie Fleming
Clusters (page 77) copyright © Dylan Dixon-O'Rourke

Also by Tim Rowe
'BEES, HIVES, HONEY! Beekeeping for Children'
ISBN 9780956702616 Green Hat Books
www.childrensbeekeeping.com

The
ROSE HIVE
Method

Tim Rowe

*To my parents, Marion and Jack,
who raised me in the country.*

With thanks to Pam Gregory for many wise words over many years.

To Karen Brock and Ed Rowe for many excellent suggestions; my beekeeping brothers Mike and Andy for reviewing the text; Ron Skingley, Lizzie Fleming and Dylan Dixon-O'Rourke for help and advice; and to Paul and Gill Smith for first giving the hives a chance.

With special thanks to Sandra, for years of patience, and to Sam and Liam, for all those labels..

lifting the crownboard

These days I keep bees simply because I like to work with them; I like the smell and the sound and movement of them. I also keep bees because they are key to the health, biodiversity and magnificence of my environment.

But I *manage* my bees in a particular way, for very specific goals, namely, health, increase and honey-production. This book explains how I go about achieving these goals using a simple hive-design which allows me to challenge traditional approaches, develop new management techniques, and above all, work more closely with my bees..

Contents:

Introduction 7

Chapters:

Why Change Things? 15
What Is A Rose Hive? 21
Healthy Bees, Unhealthy Bees 35
Using Rose Hives 51
Old Comb, New Comb 63
Swarms And Other Ways To Increase Stocks 75
Harvest 87
Winter Preparation 97

Appendices:

Hive Plans 103
Converting From Other Hives 118
Starting From Scratch 122
Resources 125

INTRODUCTION

Even though I manage around 100 hives and have been keeping bees for 35 years, it was the 'wild' colonies which really started me thinking about the way we keep our bees. Bees living the way bees have evolved to live over the last 120 million years - in a cavity made by a tree. Not that we see many bees living like this these days, of course. People have cut down most of the trees since we began farming (especially the hollow ones), so wild bees are forced into much less suitable homes like human houses. And even those are mostly empty now since the Varroa mite has arrived. So observing bees in their natural state is a rare privilege, but I suggest we could learn a lot from them - and it's high time we did.

honeybees are a woodland species - they can get all they need from mature open woodland, including a place to make their home - they have adapted remarkably well as we've transformed their world, but it seems there's a limit to what they can cope with

A hollow tree is the perfect, safe, dry, insulated home for bees, where they can build combs in the way they want to. In return, the bees deliver nutrients to the tree, collected from miles around, in the form of faeces and dead bees, presumably amounting to many tons over a few decades of occupancy. This partnership seems so beneficial to both parties that I argue it's probably not coincidence, instead, yet another example of stunning natural symbiosis.

The point is, *all hives are imitation hollow trees*. They work because they allow bees to do what bees want to do - make comb, store food, raise brood - in a breathable, water-proof, defendable space. But hives are not the same as hollow trees and they often have serious deficits - *especially* modern hives.

Modern beekeeping began in 1852 with the Reverend Langstroth's brilliant invention - The Removable Frame Hive. We all use it still, or one of it's many variations (with the exception of Top-Bar beekeepers - more on those later). It's brilliant because we can remove a comb - safe inside a wooden frame - inspect it, exchange it, extract it, and then return it, without hurting any bees. I used the National and Commercial variations of them for many years and I would be using them still, except the two sizes of boxes irritated me so much.

Brood-boxes and supers. Deep frames and shallow. Every book I read, every beekeeper I met, accepted them as standard and necessary, even though they make most beekeeping jobs complicated, clumsy and crude. With two sizes of box, the frames are only interchangeable within half the hive - a curious arrangement to say the least.

Back as a naive teenager, with my first hives and my stack of beekeeping books, I'd assumed that bees preferred hives like that - even *needed* hives like that. But since then I've seen enough of wild bees' homes to know better.

Honeybees left to themselves make combs of all shapes and sizes - from just a few centimetres tall to well over a metre. They adapt to the cavity they find, building perfect combs using nothing but their own wax secretions and tremendous skill. And then they arrange their colony within these combs in a precise but fluid way.

Bees and pheromones are constantly circulating through the combs, stores are packed and processed and re-packed, the brood-nest waxes and wanes through the seasons, and, curiously, the queen often leaves her laying to make trips out to the far corners. Whatever complex and subtle structures are involved, this behaviour has been successful for millions of years.

Even when bees were moved into hollow logs or straw skeps, clay hives, wicker baskets, fruit boxes or tea-chests, their behaviour was unaffected. They could continue to build combs - and their home within those combs - in the way they wanted. But then along came hives with two different box sizes - *and the queen-excluder which makes them practical.* Suddenly the bees' home was divided in two and I think we're only just beginning to understand the serious implications of this development.

If we add the near-constant use of imprinted wax foundation sheets, which changes the ratio of drones to workers in the colony, then we begin to see that standard beekeeping practices, the ones I was taught and the ones that are still taught decades later, are having a profound affect on our bees. I would go on from that, too, and include the attitude that permeates many books and courses, which, I'm sorry to say, amounts to a kind of arrogance. Instead of being content to watch and learn and marvel at our bees, and reap the harvest at the end of a good year, beekeepers are taught that they should try to change their bees' behaviour. In effect, that we humans know more about the best way to be a bee than the bees do themselves. This attitude extends into swarm-control, drone-production, honey-harvest, bee-breeding and bee-health and I would argue that, far from benefiting our bees, it is contributing to the massive pressures bees are under.

So that's what this book is about - addressing some of the issues around conventional hives and the way we've been taught to keep bees.

Not that I worked it all out in one day over a cup of coffee. In the beginning I was only concerned with the hives side of things. I'd expanded my beekeeping to one hundred hives and suddenly I was learning more in a season than I had in all the previous years put together. But those two box-sizes and frame-sizes really didn't make sense to me still and I was determined to find a better way. For a start, I wanted a single-size box throughout the hive that would work for both honey and brood. One that I could over-winter bees in even at one storey high.

After cutting up frames and boxes and experimenting with

many different shapes and sizes, I designed a box based on the maximum full weight I could comfortably carry. I stuck to the square shape of my standard hives so that I could continue to use all my floors, crownboards and roofs. But I chose the relatively deeper shape that a long-lugged frame allows - specifically for over-wintering in a single box.

I make all my hives, having worked out a really easy way to do it.

Immediately I knew I was onto a winner. Things began slotting into place all over. I looked at the stacks of existing standard hives in my workshop and kicked myself for not starting my experiments earlier. It took a while, but eventually every brood-box was replaced, and then every super, until I had just one size of box and one size of frame throughout all my hives and I could finally get on with sensible beekeeping. Along the way I began to see what a fundamental difference this hive was making to my beekeeping - and more importantly, to my bees.

Now this was all happening at the same time horror stories were emerging from around the world of massive colony die-offs. Even locally, many people I knew were losing all their bees. It was a worrying time for beekeepers everywhere, but my bees seemed to be doing very well. In fact, the results I was getting were very impressive - more bees, more honey than I'd ever harvested, and the beginnings of a whole new approach to beekeeping.

Having thrown out traditional hives I soon found myself challenging all sorts of traditional methods, too. And again, I found that the way we've been taught beekeeping wasn't necessarily the right way just because it had been around for a long time. One way or another, I reviewed everything I'd been taught about bees and hive-management. And I haven't looked back since. Instead I look at what the bees do - and the more I look, the more I see..

So there I was, busily experimenting with my new hives and developing different techniques, relieved that my bees were thriving, but not realising that anyone else would be interested. But they were. I wrote a couple of articles and gave a couple of talks and offered the hive-design to Thorne's Beekeepers' Suppliers - and suddenly I was getting enquiries every day. Hundreds of people started buying Rose hives from Thornes or down-loading the plans from the website and building their own. I wasn't trying to persuade anyone - only offer them an alternative, but it was gratifying to hear back from so many people who said their bees were also doing better with this new hive.

So far people in nearly 40 countries have down-loaded the plans and increasingly I'm seeing Rose hives popping up all over the place. Of course, they're only another variation on Langstroth's design and no doubt will be improved before long, but for now they're the best we have and they'll have to do.

But, as I said, the hives are only one part of the equation - there's far more to this way of beekeeping than a slightly different hive-design. Hence this little book. I'm trying to explain not just a new hive-management method but a new attitude to bees and a new approach to beekeeping - developed over many years of watching and working with bees, but clarified and solidified more recently as our bees' problems have intensified.

The hive and the methods I use really are working for me and my bees living in South-West Ireland, and I would sooner give up beekeeping than put my bees back into conventional hives. But if you're reading this in New Mexico or Slovakia or somewhere else, then perhaps some things will be a little different for you (we don't have Small-Hive-Beetle here yet, for instance). Remember, though, that you and I are working with exactly the same species of bee - they may be a different strain but their needs are identical and they have far more in common than they have differences.

I'm continuing to learn new things every time I open a hive, but in the meantime I think there's enough here to interest even the most experienced of you. I hope I have reasoned my arguments, explaining why I think this method works so well, but of course, I won't convince everyone. Luckily, that's really not the point. I'm not trying to convince anyone, just offer some suggestions and an alternative to the way we've been doing things.

I've aimed this book at people who have bees already, or at least have a good understanding of bees and beekeeping. I have not included a beginner's guide to beekeeping or an explanation of bees' behaviour, because I'm assuming you don't need those. Instead, I'm offering you what I've learned through many years of thought, accident and experiment. I hope you find some of it useful..

West Cork, Ireland, 2010

May your bees do what bees want to do.

Chapter One

WHY CHANGE THINGS?

forget-me-not, borage and bee

In recent times we're all having to wake up to the fact that bees are not quite as resilient and adaptable as we'd thought they were. They just can't cope with the endless environmental degradation that goes on, the monoculture, the pesticides, the pollution. But they also can't cope with the way we beekeepers have been treating them. And I don't just mean the big commercial bee farmers with their cattle-lorry management, I mean ordinary well-meaning beekeepers in their back-gardens with their couple of hives.

Not that anyone sets out to exploit bees or mistreat them or cause them ill-health. On the whole beekeepers are some of the nicest people around but, without meaning to, perhaps you and I have been adding to the many challenges bees have to deal with. We want to do a good job, we try hard - but, in some ways, we've failed.

In light of all the problems that bees face these days, it seems to me we need to take some responsibility for the situation ourselves and review the way we've been doing things. The way we were taught beekeeping, our hive-management, our priorities and attitude - all could use a little objective analysis. Nothing very radical but, crucially, the result could just tip the balance in favour of our bees.

Just as bees need to evolve to cope with changing circumstances, so too should beekeepers. The way we've been doing things for the last 160 years (since modern hives first appeared) has worked reasonably well, but things are changing at a tremendous pace and we need to change too, to keep up. More extreme weather conditions, the ever-increasing human population, the loss of habitat and biodiversity, the spread of genetically modified crops, are all enormous factors, and any one of them should be cause enough for concern. Hive numbers around the world keep dropping - sometimes alarmingly fast - so doing nothing, changing nothing, will inevitably lead to disaster. Steady as she goes is just not good enough. We have to change direction. Examining our own beekeeping practices is surely a good place to start.

It wouldn't matter so much, except that beekeeping is probably the single most important job on the planet! We cannot afford to fail. We are the custodians of a species that is fundamental to the life-cycles of a vast range of flowering plants, which in turn are the food for animal species big and small (- including humans, of course). If honeybees continue to decline, species everywhere will go down like dominoes. Perhaps it's our duty as well as our privilege to find a better way to work with our bees, helping them stay healthy and numerous. After all, if we don't, *then who else will?*

Of course our role is completely over-looked by governments - they still value bank-managers and barristers more highly than beekeepers, not realizing how much they need us. But perhaps that will change too, given the environmental crises unfolding around the planet. One day perhaps beekeeping will be properly recognized, and resourced and supported appropriately, but until then we'll all just have to muddle along as best we can. Helping and advising each other, learning all the time, challenging things when they need it. It's in this spirit that I offer the suggestions in this book.

For a start, beekeeping shouldn't just be about producing a surplus of honey - gratifying though that is. These days beekeeping also has to be about producing a surplus of *bees*, too - so that we can pass them on to other would-be beekeepers, rebuilding lost populations, strengthening genetic-diversity, ensuring pollinators wherever they're needed. And also, most importantly, beekeeping should be about ensuring that our bees stay healthy.

Surplus honey, surplus bees, healthy colonies - it's just common sense, I know, but sometimes it needs pointing out.

Too often beekeepers focus on the honey harvest, are put off by the complications of swarming and dividing colonies, and downright terrified by the long list of bee diseases. The results they achieve are disappointing and they end up disillusioned - and their bees often end up dead. This situation cannot continue; there's just too much at stake. We need to try a different approach.

combs brimming with honey

Perhaps we all just need reminding that these goals - surplus honey, surplus bees and healthy colonies - *are shared exactly by our bees too*. They want the same things we want - or should want. So instead of fighting them, strait-jacketing their behaviour with archaic beekeeping practices, forcing them to do things they don't want to do - we only have to relax, watch, and work with our bees, helping them and supporting them where we can. Beekeeping is easy when you work with your bees. And if we get these things right then beekeeping isn't just easy and deeply satisfying - it can be financially rewarding too. Many people want to buy healthy bees and even more people want to buy good quality honey. What other job is so interesting, so rewarding and at the same time so important?!

But that begs the question: How could so many well-meaning, hard-working beekeepers have been getting it so wrong all this time? If beekeeping is so easy, why is the bigger picture so deeply worrying?

Part of the problem is the hives we were taught to use. They're restrictive and awkward and difficult to keep clean and healthy (in some respects they're actually worse than a 17th Century straw skep - at least those were emptied out completely sometimes). The way we were taught beekeeping needs a good shaking, too (- swapping honey for sugar syrup, for instance, is short-sighted as well as greedy). There are lots of other aspects to traditional beekeeping that need challenging - but that's hardly surprising considering how little it's changed in 160 years.

When I stepped back and questioned the way I'd been taught beekeeping (which is exactly the same way most people are still taught it 35 years later) I found things I wanted to change - and when I did, my bees did better. It was as simple as that. But that doesn't mean I've come up with the definitive way to keep bees well - not at all. I'm just learning, just the same as you and everyone else is. I'll show you what I've learned and then you can tell me where I've gone wrong, and between us all we'll perhaps get a little better at this beekeeping business.

Personally, I have found that changing over to a simpler hive and altering my approach has been key to the successes I have had, and that's why I recommend this and why I'm describing my experiences here. But if you can achieve good beekeeping with your present set-up, then more power to you. Whichever way we look after our bees, let's make sure we're doing the best job we can.

Chapter Two

WHAT IS A ROSE HIVE?

Rose Hive

Rose hives are yet another variation on Langstroth's removable frame hive, first patented in 1852. So they're a lot like National hives, Smith hives, W.B.C.s and Commercials, and all the other hives you can build or buy, but they differ in one fundamental way: they are the first frame hive specifically designed to have all the boxes - and therefore all the frames inside - identical.

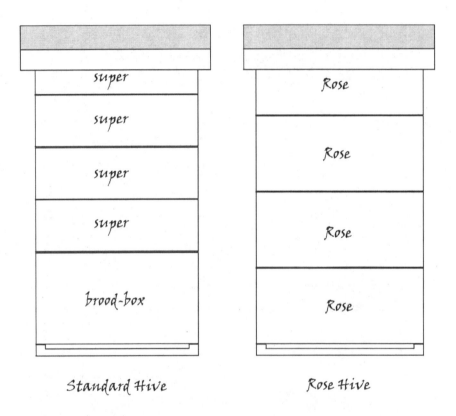

Standard Hive   Rose Hive

There are no distinct brood-boxes and no supers - each Rose box can be either, depending what's happening inside.

With only one size of box, frame and foundation throughout the hives, obviously the spare equipment also only needs to contain one size of each, too - cutting down on costs and space needed.

The boxes themselves are made very simply (having only four parts) and contain 12 frames each. This construction method makes them easy to make at home, and the cheapest wooden hive to buy on the market. *(full plans at the back..)*

Other hive parts - the crown-board, roof and floor - are the same size as those for National and Commercial hives and are fully compatible.

stack of Rose boxes - I paint my hives with 'Fence-Life' sealant (I'm not sure it does any good but it doesn't do any harm either!)

The main point of these hives, though, is the hive-management methods that can be used with them. Beekeeping really opens up when every frame throughout the whole hive is interchangeable, making lots of common beekeeping tasks simpler and quicker. This flexibility within the hive also enables a more versatile and subtle approach to beekeeping, and I'd argue leads to a more co-operative, helpful relationship with the bees.

It might seem unlikely, even far-fetched, that making such a small change in design can affect management so much. All I can say is, I was surprised, too. Honey-production, stocks-increase, and colony-health were all suddenly far simpler and more successful.

For example, because you can replace old brood-comb easily, with no waste at all *(details later)*, the entire hive is cleaner and healthier - something which is absolutely vital for the colony that lives in it. Yes, you can do this in a standard hive, but it isn't easy because there's usually brood or stores still in the combs that you want to replace, so often the job doesn't actually get done. With a Rose hive you could do it every year if you wanted to, *wasting nothing*, and it would take just a few seconds. That's not because these hives are wonderfully clever or anything - it's just that they suffer from fewer restrictions than other hives.

Don't bees need two sizes of box?

There's no biological reason at all why hives should have bigger frames in the brood-chamber than in the supers - it's just a tradition we've kept up. Wild colonies very successfully make their combs in all shapes and sizes and if bees happen to live in a hive then they really don't care what sort it is, as long as they can get on with building comb, raising brood and storing honey and pollen.

Hives were traditionally built in two sizes, not for the bees, but for the beekeeper. And specifically for the beekeeper whose main aim was honey-production. The tradition goes all the way back to when we used straw skeps, which were bell-shaped -

which meant you couldn't stack them unless the top one was smaller than the bottom one. We repeated this pattern even when we moved over to box-hives and then frame-hives and by now the design is virtually cast in stone.

Obviously, the argument runs that boxes full of honey will be too heavy if they were the same size as a brood-box, and so should be smaller. But considering that a brood-box is *invariably* too small on its own to accommodate a young queen, then where's the sense in making that brood-box the starting point in designing a hive?!

What's wrong with having two sizes of boxes?

The short answer is *quite a lot!*

All standard hives have a single brood-box that was originally designed to hold all the brood. But nowadays most beekeepers find them all far too small - any vigorous young queen would quickly run out of space to lay - so either they double up the brood-boxes, or even (as I was taught) use a brood-box and a honey-super together as the brood-chamber (the 'brood-and-a-half' system), making for really complicated beekeeping.

Having a brood chamber that isn't big enough is just plain silly. A colony whose queen is restricted is going to swarm earlier than otherwise, and it will never reach its full potential. Getting around this problem with more boxes is the obvious solution, but to use two different size boxes means the brood chamber is filled with frames you can't swap about. This is not only awkward and frustrating, it also severely limits the operations you can perform.

(The other solution - having extra-big brood-boxes - brings its own set of problems. A very large comb is likely to buckle or kink or just develop a wave through it. This wave in the comb will be echoed by the bees in the next comb. Before long all the combs are slightly different and can only fit together in a particular order - making it impossible to swap frames around effectively.)

Meanwhile, the honey supers in a standard hive were made smaller so they'd be lighter when full - perfectly sensible - except *this only works if you physically keep the queen out using a queen-excluder.* Without an excluder the two different sizes become entirely redundant and pointless because the queen will lay eggs in the supers. But using an excluder means restricting the queen's movements and forcing all the other bees to squeeze repeatedly through sharp-edged little gaps. The excluder is the most artificial part of any hive and is specifically designed to stop bees doing something that they want to do, namely deciding for themselves where to make their brood-nest (i.e., the area where they raise their brood).

The brood-nest in a wild, unrestricted colony is bubble-shaped. It changes size constantly but it never has corners. When you remember that a brood-nest has to be maintained at around 35 degrees centigrade then you can see the main reason why. The bees choose the area inside the combs that offers the best protection and the best insulation - the middle. In a colony with plenty of space, every part of the brood-nest is wrapped around with comb and bees and stores. But, when you keep a queen squashed into a square box that's too small, she will be forced to lay right out into the corners. This brood is obviously more vulnerable to extremes of hot and cold, and the workers who regulate the temperature so carefully round the clock presumably have to work much harder to keep it perfect. They also have to carry food from farther away because it's been displaced.

Having an excluder in your hive means, not just a restriction for the brood-nest, but a giant obstacle for the bees trying to store honey. Is that really what we want for our bees?

(You might be wondering how I keep the honey and the brood separate without an excluder, and I explore that in the 'Harvest' chapter. Basically I found they weren't needed at all, and the benefits they bring to beekeepers are easily out-weighed by the disadvantages they bring to bees.)

If you add in the complications to the beekeeper of having two sizes of everything, and having to keep two sizes of spares as well, then we have the obvious down-sides to the standard, conventional hive - but I would suggest there are lots more subtle ones, too, connected to the way bees arrange their home, move about within it, and interact together.

The brood-nest, specifically, is the vital core of the colony; everything depends on what happens inside it. The queen trundles around in there laying in circles and concentric rings, and the rest of the colony is arranged like an onion, in layers, around this core. Pollen in a multi-coloured rainbow around the brood, unsealed nectar and capped honey beyond that. And all the while bees moving through the combs in complicated over-lapping circuits, meeting and greeting each other, dispersing pheromones, cleaning and checking and patrolling. Whilst honeybees can - and do - adapt to most shapes of hive and cavity, they have never before, in their 120-million-year history, met a queen-excluder.

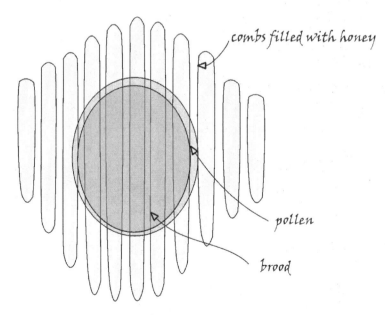

layout of broodnest in an unrestricted colony

Perhaps, instead of imposing a regime on the hive that doesn't benefit the bees at all, we should be helping them to do more of what they want to do, accepting that they probably know best what suits them. If honeybees want a bubble-shaped brood-nest arranged wherever they want in the hive, then, I say, that's what they should have.

Although the boxes and frames are what make this hive different from others, we shouldn't overlook the other parts of a Rose hive.

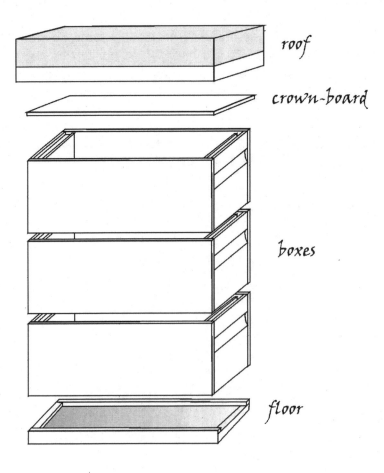

The Crown-Board

This is surprisingly important for such a flimsy component. It is essential, not so much for keeping the heat in, but for stopping draughts. An ill-fitting crown-board will allow the hive to act like a chimney, whisking away the heat. For this reason, never be tempted to make any sort of ventilation holes in the crown-board (or use one with feed-holes or one-way bee-escapes). They should always be made of thin, porous (not sealed) wood because they need to allow water-vapour to pass through easily. The water that is evaporated off the nectar as part of the transition to honey, plus the bees' own breathing, amount to many litres of water everyday in the summer, and it all has to exit the hive, most of it *through* the crown-board. If it didn't, the resulting warm fug would be perfect for moulds to grow in.

Many people make the mistake of using polystyrene as an insulation layer above the crown-board. It might keep the heat in, but it will keep the moisture in too. Layers of natural sacking or carpet, or a tray of wood-shavings or straw would work far better, but even these probably only benefit the colony in early Spring when the brood-nest is expanding and vulnerable. Remember, cold in winter is *good* for bees - it kills off some pathogens and enforces a period of rest for the queen.

(The topic of hive-ventilation is bound to raise discussion because it is quite complex. There are many who swear by through-flow systems, but really, where I live at least, the bees make their choice very clear by blocking up any hole above entrance-level. Not so long ago most floors were solid and more ventilation had to be provided one way or another higher up, but open mesh floors make this unnecessary these days, especially if you remove the solid tray beneath.)

The Roof

This fits loosely over the hive, keeping the rain out, but it doesn't sit directly on the crown-board; it's held slightly above by two thin wooden spacers - again, to let the water vapour out. It should be heavy and robust.

The Floor

The floor, too, is obviously vital. I use a varroa mesh screen nailed onto a stout frame. Adding timber battens on three sides leaves an entrance slot. The mesh provides the essential ventilation the colony needs, plus protection from invaders, and some, at least, of the varroa mites fall through. I don't use a solid sheet below the mesh, as you would get if you buy a complete floor. The bees seem to do very well with all that fresh air beneath them, remembering that there are no draughts through the hive, and my hives are on low stands and always in sheltered places.

Don't skimp on your floor - you may well be strapping up your hive and moving it to another apiary - a stout floor is what keeps it all together.

That's all there is to the hive - apart from a mouse-guard and the stand (for this I use four narrow concrete blocks, the sort builders call 'soaps').

Drawbacks

All hives have many drawbacks and Rose hives are no exception. They have significantly fewer than standard hives, though, and that's why I use them. Their single biggest downside is the extra weight of a honey-filled Rose box compared to, say, a National super. They can weigh anything up to 25kg each. That's a lot! Personally I'm delighted to be struggling along with one of these filled with honey, but they are bound to be too heavy for some people. The only way around this is to put half the frames into an empty box and move them in halves.

Weight is only likely to be a factor at the very end of the season when the boxes are filled with capped honey; the rest of the year the boxes will obviously weigh much less.

Other Hive Types

Top-bar hives are great and I would recommend them, but they are not without problems, too. For those who don't know them, they are a one-storey box with lots of parallel wooden bars along the top, each supporting a single comb. The bees expand horizontally (which is no more natural than expanding vertically), building combs on the next bar along till they reach the end. At that point you might wish you'd built a bigger hive because there's no simple way of giving them more space. (If I had to build them big enough to accommodate my biggest colonies they'd need to be huge.)

There isn't really a crown-board either so the gaps and joints between the bars all need to be filled by the bees to prevent draughts and then broken open again by the beekeeper for inspection. This is probably not such a problem in the tropics, where they are mostly used, but a significant drawback where I am.

Their combs are often beautiful, but also more fragile because they're only supported along the top edge, so moving them, even inspecting them, is much more hazardous. They often need cutting away from the sides too (don't let anyone tell you otherwise!) resulting in damage and mess. I depend on being able to make up nuclei at the drop of a hat and moving them to another apiary, or flipping (carefully) through frames looking for the queen - neither of which you can do easily with top-bar combs.

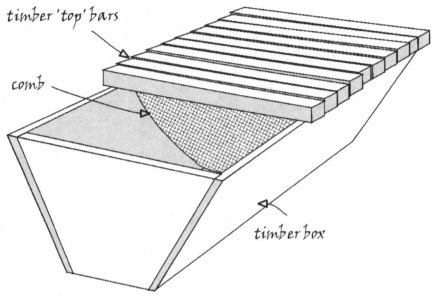

top-bar hive (roof and some bars removed)

Much is spoken about the importance of the natural comb that the bees build in top-bar hives; they can choose for themselves what proportion should be worker or drone cells. I agree completely, *but exactly the same can happen in a frame hive,* it just depends on whether (and how much) foundation sheets are used (*more on this later*).

I think we can learn a lot from the top-bar beekeepers and their approach, but if we combine this with the versatility and simplicity of Rose frame-hives we can find an even better way that suits both bees and beekeepers.

Warre hives, on the other hand, seem like a bit of nightmare to me. Essentially, they're a vertical stack of top-bar hives, but the bars have gaps between them to allow the bees to pass up and down the stack. This makes them identical to a conventional hive, except the bars don't have side-bars or bottom-bars. This means that all the advantages of having removable frames are lost completely, for no purpose that I can see at all, because without full frames the entire hive is constantly stuck up with combs.

Some proponents of both these hives claim they are more 'natural' than frame hives, but really the difference is only that the combs are not contained within a four-sided wooden frame. I can't see anything wrong at all with having these extra few pieces of timber, after all, in the wild combs are often constrained by far less benign objects. Without a frame the combs will be enlarged until they meet the sides and floor, so they end up constrained anyway. I would argue that a four-sided frame protects the comb and at the same time enables a far less damaging (and faster) way of beekeeping because the combs are not stuck to the sides.

Of course, the *management* of any type of hive can be both unnatural and detrimental to the bees inside, which is what we'll be exploring over the next few chapters..

Chapter Three

HEALTHY BEES, UNHEALTHY BEES

honeybee on purple-loosestrife

Most bees are perfectly healthy. All we have to do is help them stay that way. If we work with them, helping them do what they want to do, then they'll stay healthy entirely on their own. We don't need any pills, potions or tonics to control diseases - *bees control diseases*.

There are just two exceptions to this - **Foul-Brood** and **Varroa**.

You may be asking, But what about all the other diseases bees get? What of Nosema, Acarine, Dysentery, Chalk Brood, Sac Brood, - and all the other nasty things that fill the back of most beekeeping books and give newcomers the heebie-jeebies?

The truth is - healthy bees in healthy hives just don't get these diseases. Or, more accurately, they don't get them to the extent that they become noticeable or a problem. (I haven't seen any of these for years.)

Perhaps it would be easiest to assume that, in fact, all your colonies have all these conditions at all times. But you wouldn't know it because they're held in check by the bees. What triggers them into becoming a problem is what's important. Something weakens the bees to the point where they can't keep a lid on the infection, a tipping point is reached and the virus, or the bacterium or the parasite multiplies out of control. That's what we need to understand so we can avoid upsetting this delicate balance.

This important tipping point is easily reached when the bees are put under stress. And one way or another it's nearly always people who put bees under stress. You and I do it directly, just by being beekeepers; the changing environment we've made does it too; and we could consider the wider implications of climate change and all the rest of the changes humans have made to the world. But for the purposes of this discussion let's keep this specific to beekeepers and beekeeping..

Stress

Stress covers a broad area. I've heard people say, "But I'm always calm when I visit my bees, they can't be stressed". If only it were that simple. Even the stress of opening the hive too early in the year has been linked not only to Chilled Brood (where the brood has died of hypothermia), but also to Sac Brood and Chalk Brood. The stress caused by long-distance transportation has been linked to Colony Collapse Disorder and viral infections of all sorts. The stress of malnutrition (caused by drought, monoculture, prolonged rain, etc.) can allow Acarine and European Foul Brood to flourish, while feeding sugar syrup can trigger Dysentery. Pesticide poisoning and pollution have been linked, as far as I can see, to just about every miserable condition bees can possibly get. And then there's bad beekeeping practices like leaving dirty old comb in the hive, or taking off too much honey. When we stop to think about it, the list of stressful things we do to bees is long and complicated - no wonder they are having such a hard time.

For the most part, though, avoiding stress really shouldn't be so difficult - surely we can curb our desire to spray our bees with weedkiller and drive around with them on the roof rack? I'm being facetious, of course, but in fact we beekeepers do all put our bees through all kinds of stresses, perhaps without realizing it. Most of us avoid the obvious ones, but many of us are still doing the strangest things to our bees.

For instance, lots of people open their hives in March, mostly because all the books tell them to. March! Of course it's interesting and they may do their bees no harm, but they'll surely do them no good opening the hive so early. Spring build-up is an extremely vulnerable time for bees - they have to keep the brood-nest temperature high while outside it could be below zero. However many bees there are, they'll be stretched to do this and the last thing they need is for someone to lift the lid and let all the heat out.

(A cluster of bees is surprisingly like a baby - both of them warm, soft, but very vulnerable. All too often we don't experience bees like this because we're wearing thick gloves. I recommend you take them off whenever your bees are calm, even if it's for just a few seconds, and put your hand between two frames - I think you'll be surprised at the heat you'll feel. If you don't fancy that, at least put your bare hand on the crownboard before you open it. I suggest you avoid leaving your baby uncovered outside in March, too!)

Even the basic set up of a hive can be stressful to the bees inside - cramped space, a queen-excluder, a draughty crownboard, not enough ventilation, too much ventilation, paints and preservatives in the woodwork, moulds and bacteria in the old comb, dampness, and so on. And then there are all the curious practices that we think are normal, like feeding sugar syrup, using a smoker, even opening the hive at all. These are not normal to a colony of bees - and they're bound to be stressful.

Remember, in its natural state, a colony living for perhaps decades in a hollow tree in the middle of a wood would *never* experience any of these stresses. Nothing much would happen from one year's end till the next. And that's the way it was for all bees for millions of years. A woodland fire or a drought would be the most they might ever encounter, and then only very occasionally. (Theoretically, a colony of bees is immortal and could exist indefinitely, so, as long as a tree stands - perhaps for hundreds of years - it could be home to the same family of bees..). Compared to this, modern bees have it tough.

We can certainly make life easier for our bees by having a good look at our beekeeping techniques and our hives, removing whatever we can that might be causing them stress. Not that we can avoid everything - far from it - but if we want them to stay healthy we really do need to cut down as much as we can. Beyond the hive there will be factors we can't control like the weather and the broader environment - all the more reason, surely, to do all the things we can do.

So stress is an important factor in bee health - another is genetics.

Genetics

All the viruses and the bacteria and the parasites which, one way or another, want to eat our bees, are constantly evolving and adapting. They're trying to find a way through the bees' defences by changing slightly with each new generation. Occasionally they succeed and there's a devastating die-off, but on the whole our bees can cope with them - because bees are constantly evolving and adapting, too. And this is where we can help our bees most - and it's a great deal easier than it sounds.

Like all species, honeybees evolve a little with each new generation, through the survival of the fittest colonies and - *equally importantly* - through the deaths of the weakest. That's how they stay one step ahead of all those pathogens and it's also how they cope with gradual changes in the climate or the environment around them.

But honeybees can only make genetic changes at the moment an egg is fertilized, bringing together genes from the drone and the queen. Every time this happens an exciting new combination of genes can occur and some of the offspring will be better suited to their new challenges than either parent. That's how evolution works.

But, unlike mammals, for instance, the only fertilized eggs that count are the ones that develop into queens - the others all become workers which are infertile. However tough and resistant they are, worker-bees will never be able to pass on their genes directly to the next generation - only the drones and the queens can do that. Drones, though, get all their genetic material from their mothers. (They don't have any fathers at all.) They are made from *unfertilized* eggs, so there's no chance of mixing genes up when they are made. (Although drones don't have fathers they do have one grandfather!) What this means is that, in a way, everything depends on the queens.

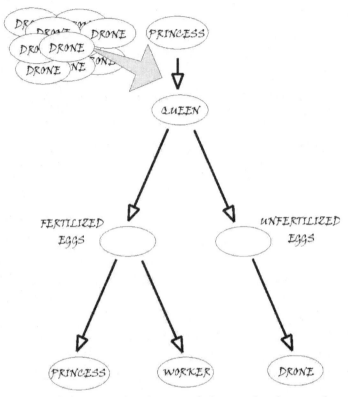

provided the queen has been mated properly, she can choose whether to lay a fertilized or an un-fertilized egg

Now, given that most queens never live long enough to lay eggs, and that most drones never get to mate, that leaves only a tiny number of bees to carry on those vital genes into the next generation. Perhaps just one queen a year from your hive will successfully take over the colony or head up a new one. That means that, while the viruses and bacteria make new generations of themselves in just a few hours, your bees might only manage one new genetic generation in a whole year.

young queen

With genetic change so important, yet so infrequent, we can see how much depends on each new successful queen - *and on how well she gets mated.*

A princess may mate with up to forty drones and carry their sperm with her for the rest of her life. Most of that sperm will be used for making worker bees, but not all; the sperm from one of those drones will fertilize the egg which one day will replace her as queen. Any one of those forty drones could be the one, so it's vital that they are all fit and strong.

All the available drones in the neighbourhood congregate on warm, still days in the summer, to cruise around hoping a virgin princess will come by. Most days they go home disappointed, but occasionally a newly-hatched princess will fly through this column of drones on her mating flight. Immediately all the drones set off after her and the fittest and strongest get to mate (whereupon they die). You know this already, I'm sure, but the thing to remember is that it's the *size* and *make-up* of this gang of drones that's crucial - the number of drones that have collected together from all the hives around - because the top forty drones out of a large, varied group will be, on average, far better than the top forty out of a small group of related drones.

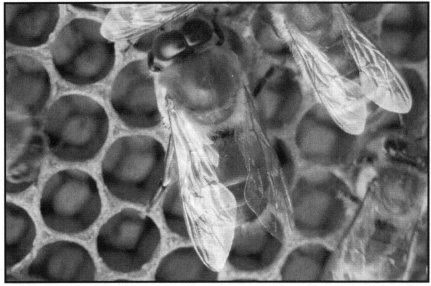

drone

Although this may seem a rather precarious system it has obviously worked very well so far, and for all their history bees have managed to cope with changing landscapes and climates, as well as constantly-evolving pathogens - but in recent times people have radically changed things for bees. Before farming became widespread across Europe there would have been honeybee colonies living in hollow trees every few hundred metres from the Ural mountains to the Atlantic Ocean. The genetic diversity within the bee population would have been *immense*. Thousands of drones from dozens of different colonies would have been chasing each virgin princess on her mating flight, ensuring, if not automatic health, then at least a healthy mixture of genes.

These days the population of honeybees is a tiny fraction of what it once was - which has meant a collapse in the gene-pool. For instance, here in South-West Ireland, I estimate (because there are no official figures) we've lost 95% of our honeybees over the last 30 years. This situation has been echoed right around the world, as you know (although not always as severe) and it has very serious implications for the genetic diversity within the remaining bee population.

With fewer colonies around there are obviously fewer colonies represented at each mating. This is bad enough, but *the way we keep our bees has made things even harder for them*. We've been severely restricting the number of drones made in each hive with our beekeeping methods (see 'Combs'). This has meant fewer drones available for each mating. So it's no wonder our bees are struggling - *we've been stifling their best efforts to stay healthy!* Bees need to evolve to keep up with the pathogens and their environment, they do their best to evolve - but we beekeepers seem to have been doing our best to stop them!

Sad to say, this worrying situation has been made even worse by people buying in queens from big commercial queen-breeders. Genetic diversity depends on having many different families or individuals within a population, each with a different lineage or history. But commercial queen-breeders produce queens that essentially all belong to the same family. And every year the number of international breeders goes down whilst the number of queens they raise goes up - a sure recipe for further shrinkage of the world-wide honeybee gene-pool.

The Solution

All this genetics business is fascinating, but a little complicated, and it puts some people off, so let's cut to the chase. The simplest way to help your bees stay healthy is to raise more colonies, and within each one, raise more drones. In the long run, more bees means healthier bees, and that's where we can help so easily. Again, we just have to work with our bees because that's exactly what they want to do, too.

Raising extra drones, queens and colonies is such an interesting aspect to beekeeping - but lots of people avoid it, spending a lot of time and effort needlessly trying to stop swarms, for instance. But when we remember that having extra colonies around is not just essential for the good of our environment, it's also essential for the on-going health of our bees, then we can see why it should be one of the main aims of all of us beekeepers. That's where the Rose hives come in - they make the job of raising more stocks very easy indeed.. *(see 'Swarms')*

Mind you, it's not much good just expanding our stocks and then buying in queens from somewhere else (whether from a big commercial queen-breeder or not). All we get that way is someone else's experiments. The bees in your local area have spent countless generations evolving and adapting to your unique local conditions - gathering an absolutely priceless genetic make-up - they're the ones you really should be breeding from.

If you import queens you'll immediately dilute your local strain and it could take many decades to undo the damage. Not only that, but any farmed queens will be 'emergency' queens (that's how they're made in such numbers) and you can do much better yourself. Plus, any bought-in queen will probably have been mated artificially, too - a method which can have serious drawbacks.

Artificial mating (insemination) involves *people*, not bees, choosing the origin, the number and the individual drones involved. There's no high-speed chase through the tree tops to sort out the wheat from the chaff, no furious life-and-death struggle to be the best. Instead the selected drones get unceremonially squashed into a petri-dish, and the resultant mess is stirred about a bit and then injected into the hapless, upturned princess.

Increasingly this is the way bees are bred around the world and it surely is part of the problem bees face. We can do our bit to stop this nonsense by breeding our own queens - and they'll be far better queens than we can buy in anyway.

Colony Deaths

The other side of the genetics coin is that some colonies *don't* have what it takes. They aren't strong enough to resist the latest virus or bacteria mutation, and they become weak and die. *That's ok too.* It may sound callous but that's what's supposed to happen. If you try to prop up a colony with antibiotics or tonics or supplements (people do), you're not only postponing the inevitable, you're making things worse for all the other bees around. If you have a sickly colony, or one that just doesn't thrive, by all means try re-queening it (this may well work), but otherwise let it die. Sad, yes, but necessary for the continuing health of all your other bees - and honeybees in general.

When (not *if*) you lose colonies, take heart; the ones left alive are the only ones you want around anyway. I expect to lose between 0 and 20% of my colonies every winter. I used to feel bad about it, feeling responsible, but now I'm actually quite pleased because the ones left alive are already stronger than last year's average. I can easily make up the numbers again during the summer - and I'll be breeding from survivors.

(I should also mention perhaps that the genes in bees are also triggered sometimes from dormancy into action. This allows a useful thing called Adaptation, meaning they can cope with changes in their circumstances faster than the evolutionary process on its own would allow. But of course, the genes still have to be there in the first place. In the long-run, unless the environment stays exactly the same, all organisms need to evolve to keep up.)

In Summary

You can help your bees stay perfectly healthy by actively pursuing these measures..

Hives - keep them dry and draught-free, *and replace dirty old combs regularly*. Fortunately, this is a far simpler job with Rose hives than in any other.

Breeding - breed lots, but don't buy in other people's mass-produced queens. Instead, make the most of your local bees, those that have adapted over countless generations to the climate, diseases and habitat which are specific to your area. Breed from your best queens and ensure all your hives are full of drones. Over time you really can improve their disease-resistance by a mixture of selective breeding and natural selection. Again, with this hive, this is really easy - and truly fascinating.

Environment - this also is key to the health of your bees. Without people the countryside around you would naturally revert to being perfect for bees - open mixed woodland, with a succession of flowers from early spring to late autumn. So it's people that have made it less than perfect - with our farming practices, our houses, cars and chemicals - and it's people you need to influence to improve the environment for your bees.

This is not as impossible as it sounds. Go into your local schools with a smoker and a bee-suit and a frame of honey. Explain what you do and why you do it. Give talks to local groups, put up photos in your local libraries, hand out willow slips for planting - anything to raise awareness amongst the farmers and gardeners around you. If you don't feel confident or knowledgeable enough to do this on your own, then assist your local association, or just take in one of the many excellent films that have been made about the importance and plight of honeybees. In my experience, everyone loves to see a hive and try on a bee-suit and taste honey directly from the comb. You'll make friends and, more importantly, win friends for your bees.

Honey - leave them all the honey they can eat for the winter. *(See 'Winter Preparation')*

Your bees will thrive with this kind of attention and you can cheerfully forget all about the diseases other people worry about.

Now, back to those two important exceptions:
Foul-Brood and Varroa.

Foul-Brood (American and European)

Theoretically, resistance to the bacteria which cause these diseases could be improved, too. After all, they can't be fatal to all bees or there would be no bees left. But that is not an option because they are both <u>notifiable diseases</u>. All stocks that have AFB <u>must be destroyed by fire</u>, and those with EFB must be treated under strict direction of the relevant authority in your country. I understand that AFB is generally rare now, but we still have plenty of it here in Ireland - mostly because we have no system of bee-inspectors and the requirements surrounding it are not enforced. Burning bees is, without doubt, the most miserable part of beekeeping, but if we all did it we could eradicate the disease altogether.

Not surprisingly, the occurrence of both these diseases has also been linked to stress. So even if they are in your area your bees will probably be able to keep them at bay while they are fit and strong and stress-free.

If you live in a country that takes these diseases seriously and offers a monitoring service to beekeepers, then count yourself lucky and make the most of them. If not then you'll have to rely on your local beekeepers to inform themselves and act responsibly. *(See 'Resources' for what to do with suspect samples of comb.)*

Varroa

Resistance to the Varroa mite in the European honeybee is, as far as I can see, only a distant hope. Of course that would be the perfect solution - to breed a bee that can handle this foreign parasite all on its own. But resistance is going to take a long time to develop, if it ever does, and until then we have to use other methods to control its numbers within a hive.

(Interestingly, it seems that it's the mite that could be developing a more sustainable relationship to the bee, and not the other way round. Seemingly, the mites in some wild colonies don't always multiply to fatal levels. Much more work is needed to find out what's really happening, though.)

Even if resistance was bred into bees somewhere in the world, we should all be extremely hesitant to rush out and replace our bees with this new strain - the honeybee gene-pool would shrink to almost nothing, leaving the bees hopelessly vulnerable to the next big problem. In fact, here in Ireland we have a permanent ban on the importation of bees and queens, put in place to protect our native Irish black bees which were being swamped by 'Italian' strains, and I say long may it remain.

So Varroa mites will remain a challenge for the foreseeable future at least, *but we can easily meet that challenge.* We just have to be vigilant, looking for mites or the symptoms of mites, and then using one of the many treatments available. I use different things in different apiaries - partly to compare the results and partly to spread the risk of anything going wrong.

Whichever monitoring system and treatment you choose, remember some treatments are not an option until after the honey has been removed, and Oxalic and other acids are lethal to bees in the wrong dosage. *(see 'Resources')*

Only treat against Varroa when there is a problem - every type of treatment has a stressful and detrimental effect on the bees so should never be used unnecessarily. And, who knows, perhaps your bees are the ones with some resistance - if you treat routinely you'll never know.

Varroa mites undoubtedly make bees less able to cope with other diseases. So you may find that your bees died of, say, Acute Israeli Paralysis Syndrome (which is caused by a virus) but the real cause is still Varroa.

healthy brood in clean comb - the goal of every good beekeeper

Health is a broad subject and in a way *everything* in this book relates to it: all beekeeping practices should have health at their core. Breeding, hives, combs, nuclei, over-wintering - one way or another everything has an impact on the overall health of your bees. So, although the headings change, we never really leave this subject..

Chapter Four

WORKING WITH ROSE HIVES

Standard conventional hives are so restrictive that beekeepers using them are sometimes reduced to a policy of adding empty supers in spring and hoping they'll be filled by autumn. But with a Rose hive all sorts of other possibilities arise. We're able to respond to the needs of our bees with far more flexibility (for example, giving them space where they need it), and the results can be very satisfying. When we allow and encourage the bees to do what they want to do we'll be rewarded by rapidly-growing colonies, healthy bees and lots of honey (weather permitting, of course).

On a very simplistic level, hive management is about adding and removing boxes. But knowing when to do this, and whereabouts on the stack to add your boxes, and changing the order of boxes within the stack, are all crucial to helping your bees. I'm offering a simple overview here, but I'm sure that you'll find many other opportunities to experiment within your own hives, given that any box can be put anywhere, and any frame will fit anywhere within them.

Adding boxes

In the spring, your bees will be expanding their brood-nest as rapidly as the weather, and their numbers, allow. Sooner or later they will run out of space and need an extra box. If you put a box on top of the hive they will eventually expand the brood-nest into it, but it will take them quite a while. This is because they will have to dig out all the honey and pollen from around and above the original brood-nest and then repack it in the new box. If the honey is crystallized, this job will be much harder and take longer. They will never leap-frog honey stores and extend the brood-nest beyond it - the brood-nest is always maintained as an integrated whole.

So when they need an extra box I always put it *between* two occupied boxes..

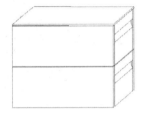

Over-wintered hive

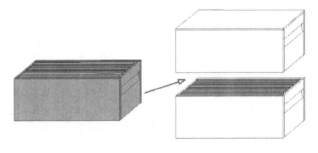

Adding an empty box

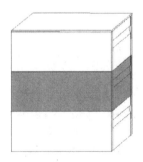

Expanded hive

Putting the new box in the middle of the brood-nest means the bees immediately have more space for brood-raising. This encourages fast build-up of the colony. It is extremely effective as long as you only do it when the bees need it; the new box must only be put between two boxes that are full, or nearly full. With experience you will quickly learn whether your hive is ready for more space. Lever open the hive in the middle like a suitcase - if it's teeming on both exposed faces go ahead with expansion. You often won't need to pull out any frames at all.

this hive is ready for another box

Inside, the bees will quickly fill up the new space, rejoining the two halves of the brood-nest, building new comb if necessary or repairing and cleaning old comb.

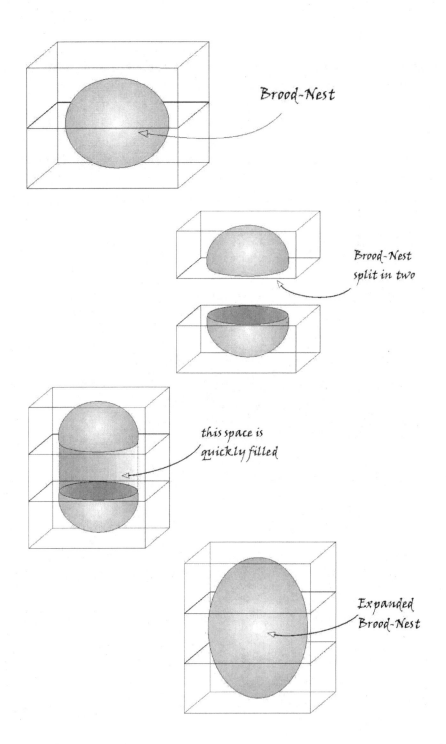

The queen will lay eggs in large, even blocks and the total brood-nest could double in a week..

a lovely frame of capped brood

I have been using this method (spreading the brood vertically) for many years now and have never had a problem with chilled brood. I assume that this is because the brood is always sandwiched between full combs with bees on them, and that the heat is retained where it's needed. The other method of spreading the brood, i.e., horizontally (where empty combs are put between brood-combs in the same box) can sometimes lead to chilling in cold weather.

Adding more boxes.

Your bees' priority is brood-expansion right into June. You can help them by continuing to give them room as they need it in the brood-nest..

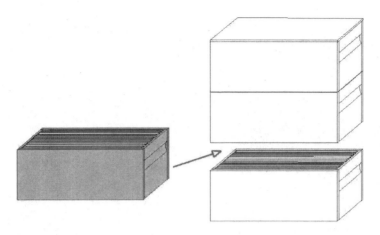

Add more boxes, as needed, into the brood-nest.

Remember, only add another box when they've used up the space they have and are ready for more. Depending on the weather and their queen, they may need as little as a week or as much as month between additions - just inspect them regularly and be ready to act when they need it.

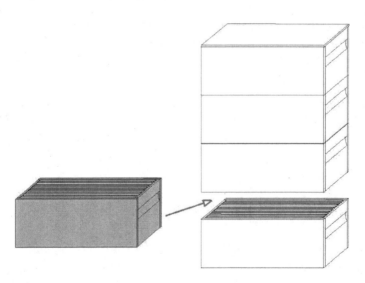

Continue until the middle of June

After the middle of June..

Generally, bees stop expanding the brood-nest sometime around mid-summer (21st June) - which means the biggest population of adult bees will arrive three weeks later, when they hatch in early July.

Depending on the weather and the flowers available in your area, this could coincide with massive nectar-flows from blackberry and clover. Or it could equally coincide with endless thunderstorms or drought. If the timing is right, boxes of honey will be saved in a staggering frenzy of activity; the bees will work ceaselessly not just during daylight hours but all through the night, too. (A midnight walk through your apiary at this time of the year will be rewarded by a wonderful deep rumbling sound from the hives - countless bees are creating the warm draught with their wings that evaporates water from the nectar. The smell, too, is just glorious.)

On the other hand, extended rain will force the frustrated bees to stay inside where will they eat into their stores. Our native bees will fly in almost any weather, including heavy rain, but the flowers they can feed on are limited to those that stay dry, like wild fuchsia and borage which keep their heads down.

Assuming at least some good weather, we the beekeepers need to keep adding more boxes as honey comes in. This time, though, the bees don't need the extra space in the brood-nest, *they need it just above it..*

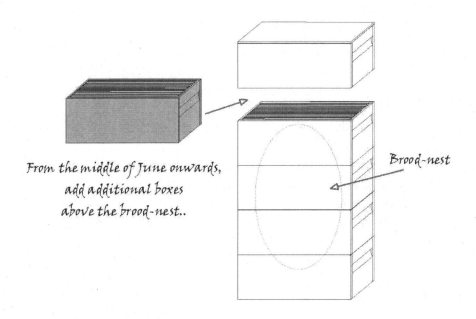

From the middle of June onwards, add additional boxes above the brood-nest..

Brood-nest

Now, the place to put extra boxes is not always clear, because at this stage the brood-nest is shrinking fast. As brood hatches, the comb could have eggs laid in it again, or it could be filled immediately with nectar, either fresh or at any point along the transition to honey. Again, the aim here is to work with our bees - they want to use the space made at the edge of the shrinking brood-nest as the first place to put incoming nectar. They pack it; evaporate water from its surface; unpack it, mixing in enzymes from their mouths and honey-stomachs; then repack it. This process is repeated until the initial nectar has become honey that's concentrated enough to be stable. If we make sure they don't run out of space in the centre of the hive, the bees will concentrate this nectar-processing in this one place, between the brood-nest below and the sealed honey above, finally packing the finished honey up against the existing sealed honey at the top of the hive.

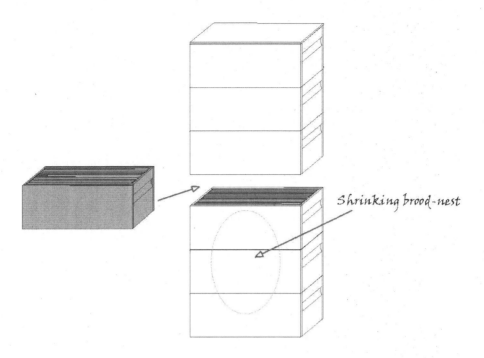

Shrinking brood-nest

Luckily, placement of the boxes at this stage isn't as critical as it was earlier in the year. I aim to put the additional boxes just above the brood-nest but as long as they are below the capped honey, they'll be fine.

You could put empty boxes on top of the stack in the conventional way, but if you do, the bees will have to carry everything through the existing honey stores. They may even unpack and move sealed honey up there to avoid having to do this. They also would have farther to move the draught of warm air that evaporates the nectar.)

Queen excluders

You will notice that during this hive management section there has been no mention of a queen excluder. That's because with Rose hives they really aren't needed at all. This means the queen can move unrestricted through the whole hive and the brood-nest can grow as big as is needed. As I pointed out earlier, excluders are not only unnecessary, they hamper the movement of the bees and restrict the queen. Their sharp edges also physically damage the bees (though the wire ones are better). I never use them anymore and my hives are the best they've ever been. But it's up to you; use them if you want to, only do try one hive at least without one to see the difference.

Many non-beekeeping people ask me what stops brood and honey getting mixed up on the comb if I don't use an excluder. And even, How do I avoid putting brood in the extractor? I explain to them that brood is supposed to be surrounded by honey and pollen stores, *that's how the larvae get fed,* so of course there will be combs that contain both stores and brood. But that's a good thing, it's the way things are supposed to be in the hive.

And as for taking brood out and mistaking it for honey - it simply could not happen to anyone who knows the first thing about bees and honey. As we all know, the wax cappings on honey are crinkly and brilliant white - utterly distinctive from the pale brown domed cappings of brood. *(more on this in the Harvest chapter..)*

Given the freedom within the hive to arrange their brood and stores as they will, bees shrink the brood-nest down towards the bottom of the hive in the late summer. This is something the bees naturally do on their own, but it makes life very easy for the beekeeper, because the surplus honey is all safely above the brood - naturally separated *without the use of an excluder.* (Of course, there will still be honey around, and sometimes even below, the brood-nest, but I don't count that as *surplus* honey.)

Occasionally I have found brood-nests which are not so much onion-shaped (the normal shape, as far as I can see), but rather more shallot-shaped. That is, they taper up through the centre of the hive, running almost to the top. Even this is temporary and within a few more weeks the brood-nests of even the biggest colonies are easily contained within the bottom two or three boxes.

So that's how you might add and remove boxes during the season on an uncomplicated hive - one that is used for honey-production. I would expect only around half of my hives to behave like this every year because they have new (or late-last-season) queens in them and all they want to do is expand and store honey. Many of the others hives need different care because at some point through the season they will have begun swarm-preparation or because they are nuclei, or queen-rearing colonies, or queen-mating colonies. But the basic principle remains throughout: I give the bees space *where* they need it, *when* they need it.

Two last points..

This way of adding boxes where the bees need the space will result in wonderful big hives - spectacular, even - and prove just how successful a colony of bees can be when given a good hive. They'll build up much faster than in a conventional hive and get impressively big. (I have to stop when the stack reaches eight boxes high because I can't reach any higher.) *But* if you feel daunted by a hive that's both bigger and heavier than you by far, then just don't let them get so big. You're in control. Slow them up by putting extra boxes on the top instead of in the middle, or, better still, divide them up and make an extra hive or two (or six!). Remember, beekeeping is supposed to be pleasurable, if you find yourself intimidated by your bees, then change the way you manage them.

Some people will question how natural it is to plonk a new box in the centre of a hive at all, but in fact, in the wild, colonies do something that's not so very different. They don't simply build extra comb up and up indefinitely (or down, either) - they add comb to the outside in any available direction, overlapping some older comb, but not all. This means they don't (or rarely) have to move capped stores, and they can expand their brood-nest sideways as easily as upwards.

There is no practical hive that allows this amount of freedom for the bees - top-bar hives force the bees to build sideways (through capped stores), conventional hives force them to build upwards, (again, through capped stores) - but with a Rose hive they can build immediately in the middle. Not the same as a hollow tree, I admit, but perhaps the next best thing.

Chapter Five

NEW COMB, OLD COMB

Comb is the backbone of the whole colony, providing structure, insulation, storage, protection, and a nursery - but often beekeepers pay it little attention. This is a serious mistake!

Old comb

One of our most important jobs as beekeepers is to throw out old comb so the whole hive stays clean. We have to do this because when we moved bees into hives we interfered with the bees' relationship with wax-moths. Wax-moths get very bad press, which is a shame because they have been an essential part of the honeybee story for millions of years; without them honey bees would have died out long ago because wax-moths are one of the very few animals that can digest wax.

In their real home - hollow trees - honeybees build new comb every year, *and then deliberately abandon old comb.* Whole sections of comb are left unattended, however big the colony is, while they extend their home in a different direction. Along come the wax-moth caterpillars, like a team of demolishers, and tidy all the old, dirty, diseased comb away. The bees have an clean empty space to build in next year, while the fat little caterpillars pupate in the debris on the floor. A neat solution and both sides benefit.

In the hive, however, there simply is not the room for bees to do this. They are forced to continue using combs riddled with bacteria and moulds - no wonder colonies get sick. Bees constantly clean and tidy, they use propolis everywhere as an antibiotic and fungicide, but there comes a point when old comb is no longer safe to have around. Brood-comb - where vulnerable larvae live - is particularly dangerous because each generation of hatching bee leaves behind an extra silken cocoon in the cell, each one stuck to the last, trapping faeces and microbes.

I have seen brood-comb as thick as cardboard and literally as black as tar, possibly decades old, in the brood-boxes of conventional hives - and then people wonder that they have chalk-brood or dysentery, or whatever. A far cry indeed from the sterile, delicate, translucent white comb that young brood should be raised in.

Comb replacement is where a great many beekeepers fall short. They leave old comb in their hives and the bees have no choice but to put up with it. Diseases of all sorts build up and eventually even the strongest colonies will get sick. The answer is simple - act like a wax-moth and remove old comb, giving the bees room to build new, clean comb. Everyone knows they're supposed to do this, but the truth is, in a conventional hive it's not that easy and the job often doesn't actually get done.

Even the modest ambition of replacing a third of your brood-combs every year is difficult in a conventional hive - because it would be very unusual to find four empty combs in the brood-box of any reasonably sized colony. No one wants to throw out comb that has brood or honey in it - however manky that comb is. But to have a chance of finding empty combs you'd have to open your hive very early in the year, before the brood-nest has expanded far - not something anyone should be doing.

With Rose hives, however, the job is done in a very different way and it's very easy. Using the expansion methods from the last chapter, a box that contains old comb can be gradually worked up to the top of a Rose hive over a season and then removed. On it's journey upwards the bees will naturally replace the brood with honey - which can be extracted before the comb is finally rendered down. This method means there is no waste at all.

Sometime in the spring, I switch round the boxes in the hive-stack so the bottom one is no longer at the bottom..

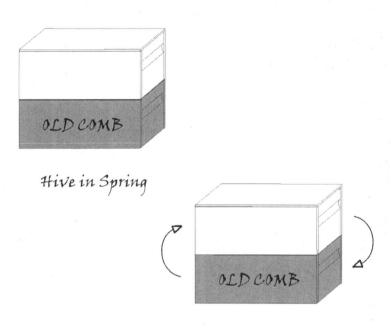

Hive in Spring

Swap around the bottom boxes

As additional boxes are added through the season..

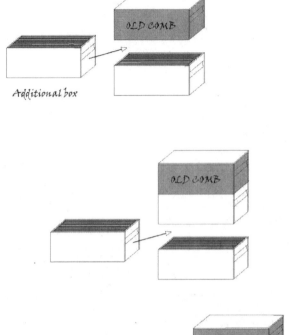

… the box with old comb inside works its way to the top.

Meanwhile the brood has been replaced by honey, ...

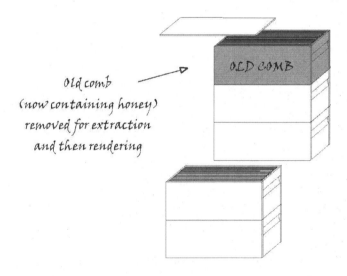

Old comb
(now containing honey)
removed for extraction
and then rendering

.. which can be removed for extraction, leaving the old comb empty and ready for rendering down.

Swapping round the bottom boxes every year or two, combined with this expansion method, means that all my comb can be emptied and inspected regularly, *and it only takes a few seconds.* There are no combs older than three years in any of my hives and I am sure that is part of the reason my bees do so well.

After the frames come out of the extractor they either go back into the box (if the comb is in good condition), or onto a pile that ends up in a large tank of boiling water. After a few minutes in the tank everything melts away from the frame. The clean, recovered wax can be re-used, so too the frames which are now sterilized (at least to some extent). The thick sludge of dirt, old pollen and cocoons that remains in the tank ends up on the compost heap. The water itself has now become a rich wine-must and can be fermented. Even the wire pieces from the original foundation are recoverable..

This is a satisfying, if messy, job and results in a stack of clean frames, blocks of sweet-smelling wax and lots of bubbling wine-buckets. More importantly, it is a corner-stone of my bees' health policy.

blocks of rendered wax

New combs

Now this is where we have a chance to influence the way our whole hive is structured, which has wide-reaching implications.

As you know, combs are made up of worker cells or drone cells, or a combination of the two. When we remember that the proportion of worker cells to drone cells helps determines the proportion of adult workers to adult drones, then we can see how important the make up of combs is to the overall balance within the colony.

In a modern hive this balance has been skewed by the near-constant use of worker-cell foundation sheets. These wax sheets (invented by Johannes Mehring in 1857) have been rolled with a die which imprints indentations in the wax which are the exact size and shape of the bases of worker-cells. This doesn't mean that the bees *have* to build worker cells on the sheets, but it certainly encourages them, and that's mostly what they do. The result is lots of worker-comb and not enough drone-comb in the hive - *so there aren't enough drones*.

Traditionally, drones were seen as almost unnecessary and drone-comb was a waste of space. That was back in the day when it was thought that the virgin princesses mated with just one drone, and very few people recognised the importance of genetic diversity.

Now, though, we're beginning to understand the role and importance of drones and it seems that, funnily enough, bees had it right all along. When it comes to the mating of princesses, the more suitors she has the better. This competition on the mating flights, with drones vying with each other for the privilege of mating, is *the only way bees have of maintaining good genetic health.*

The quality of the queen's offspring for the remainder of her life, and specifically the genetic make-up of all the queens she might produce, hinges on those few minutes she spends in the air with the neighbourhood drones. It's hard to imagine, but the consequences of those ten minutes, good or bad, will ripple down for all the years to come. Restricting the number of drones available - *which is exactly what beekeepers have been doing for generations* - must surely have had a negative effect on queen-mating.

We might all wonder what damage has been done inadvertently to the genetic make-up of our bees by generations of foundation-wielding beekeepers. It seems likely that we have influenced mating conditions to the detriment of the entire honeybee population, and that's not even taking into account artificial insemination or intensive breeding at all. Yes, there would have been drones from 'wild' colonies for most of this time, contributing to the drone-pack, but now even they are gone.

So I say it's high time we looked again at our own hives and asked ourselves what we can do to help the bees stay genetically healthy and diverse - and the simple answer is to allow them to produce more drones. Which brings us back to combs..

You will have noticed, I am sure, that any damaged comb in your hives will be repaired with drone-cells, and any gap made by a missing frame will be filled with drone-comb. Given any opportunity the bees will build drone-comb in preference to worker-comb. Clearly, they are telling us as loudly as they can that the balance is not correct in a standard hive. Given the freedom to do whatever they like, bees would undoubtedly get the balance just right, so let's allow them to get on with it.

In practice, though, if we let bees build combs freely in the hive they might have all the drones they want but we wouldn't be able to get any frames out at all because the combs would be all over the place. Whilst bees build beautiful, even, parallel combs with no help from anyone, they will happily build them diagonally across the hive, or change direction halfway across, or start on the wall, or adopt some other pattern that's equally inconvenient for us beekeepers. The trick is to get them to build the comb they want, but build it inside the frames.

Top-Bar beekeepers use a line of wax, or a narrow strip of foundation along the underside of the bar to get the bees started in the right place. After that, the bees will build comb with whatever proportion of worker to drone cells they choose, and the comb will hang down evenly from the top bar (mostly!). This is called free comb and couldn't be bettered by man or machine.

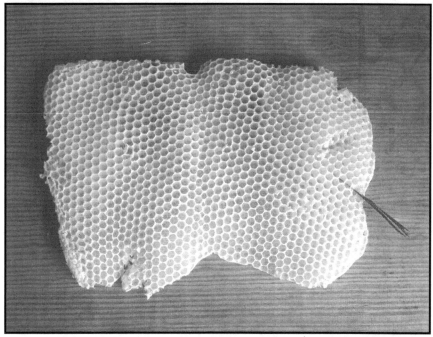

'free' comb built entirely by the bees

But we, too, can do exactly the same thing in our frame hives. Remember, a frame is only a top-bar with sides and bottom. Those four extra wooden pieces that make up a frame stop the comb getting stuck to the sides and floor (a very good thing too, in my opinion) and make the whole thing more robust - but they don't influence the structure or the make-up of the actual comb inside.

So, if you like, you could have free-comb throughout your hives. Just put a thin strip of foundation into each frame and the bees will do the rest..

frame with strip of foundation fixed along the top..

*..the same frame a few days later -
usually they go on to fill the frame completely but
sometimes they leave out the bottom corners*

Personally, I prefer a compromise. I use just a strip of foundation in many of my frames, but I still use whole sheets too because it helps maintain flat, even combs throughout the box - essential if you want to interchange frames freely. Perhaps one-in-three of my newly-refurbished frames will contain just a strip of wax, the others get the full sheet of wired foundation. The free comb they build in my hives contains lots of drone-comb, but not all, so I assume they are happy with the ratio they achieve. This is an area I'm constantly reviewing, though, because I am convinced it is vital for my bees' long-term success. If they want more drone-comb then that's what they'll get..

Some will say I'm only encouraging Varroa mites with all this drone-comb (the mites prefer drone cells to lay in because they can squeeze in another generation before the drone hatches), but I say tackle Varroa with different strategies. *We need more drones around.* If you limit the number of drone cells in your hive then they will probably all be attacked by Varroa mites anyway. This would be disastrous because you may have no fit, flying drones in your hive at all.

There's another reason for having more drones around - curiously, but unmistakably, a hive filled with drones is a calmer, more productive, healthier place..

One last point..

When a new box is put into the middle of a large hive, comb-building can be very fast and furious - so fast and furious, in fact, that the heat generated can cause un-wired full sheets of foundation to sag under the weight of bees before the comb is completed. That's why I use wired foundation if I use a full sheet. (It also stands up much better in the extractor than does free-comb.)

Chapter Six

SWARMS AND OTHER WAYS TO INCREASE STOCKS

For all sorts of reasons it makes sense to increase our stocks of bees every year. Not just to replace bees that might die over winter, not just to be able to sell or give away a few nuclei, but also to increase the genetic diversity of our bees. Better, by far, to have more stocks than you need and choose the best to keep, than have the bare minimum and risk disaster.

Fortunately, bees also want to divide and sub-divide every year, so it's no surprise that making increase can be easy. And especially so with these hives. It's also very satisfying. I try to follow the lead of the bees, work with them when I can and keep things very simple.

Left to themselves, as you know, bees make increase by swarming. On the whole this method works extremely well and we could learn a lot from it. The swarming colony prepares very carefully. It only swarms when it's in prime condition, it always leaves a well-fed princess behind to replace the queen (- and plenty of back-up ones, too), and those leaving are fit and well and absolutely stuffed full of food. Both the half that stays behind and the half that leaves stand a very good chance of success.

a fine big cluster

A swarm is a noisy whirling cloud of thousands of bees, darkening the sky and filling the air - completely beyond the control of humans. But, within minutes, that tornado transforms imperceptibly into a quiet shimmering cluster hanging from a bush; heavy, warm and literally vibrating with energy and anticipation. Humans have been catching these clusters for thousands of years and when we do, and pop them into a hive, then we really see the potential of a colony of bees. A good-sized swarm early in the year will work harder and faster than any other colony, building comb, raising brood and storing honey in an enthusiastic race to be ready for winter.

clusters are warm and silky-smooth

Now, the one snag with swarms is that, although they give us plenty of warning (swarm-cells), and only choose nice afternoons to appear, and they are noisy and highly visible, and they give us every chance to catch them by hanging around for hours (even days) on a nearby bush, - still we often manage to miss them. They end up in some inaccessible hole with nobody to watch the Varroa for them. (Or worse, they end up in a neighbour's rotten hive and still nobody watches the Varroa for them..!)

The half that's left behind may also complicate things by casting more swarms, this time each headed up by a virgin princess. These are far smaller and more vulnerable and less likely to succeed, but still the bees do it, presumably because a slim chance of increase and genetic diversity is better than none at all.

Catching clusters was for a long time our only method of increase, but it has fallen out of favour and now all the beekeeping books teach something called 'swarm-control'. In its crudest form this usually amounts to checking through the hive every nine days and squashing any swarm-cells you find. Apart from the shocking waste involved, *this method is almost bound to fail sooner or later.* All you have to do is miss one weedy, crooked little swarm-cell hidden in a crevice and your management plan is blown apart.

I do nine-day checks, too, looking for queen-cells, but my aim is entirely different. If I find them I know the colony is ready and preparing for division. I take my cue from them, dividing colonies in different ways, working with my bees, confident that the timing is spot on.

Now, any successful new hive or nucleus has to have two things - plenty of worker/drone bees and a queen. Let's take the bees part first..

A Rose hive can be divided ridiculously easily into as many parts as you like. You could simply divide your hive in half, providing an extra crownboard, roof and floor..

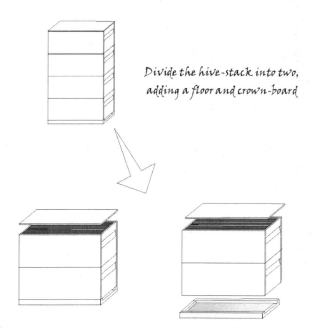

Divide the hive-stack into two, adding a floor and crown-board

You could take each individual box away separately..

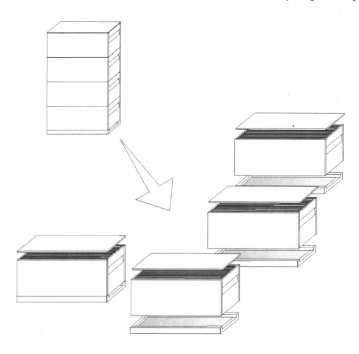

You could fill nuc-boxes with frames from anywhere within the stack..

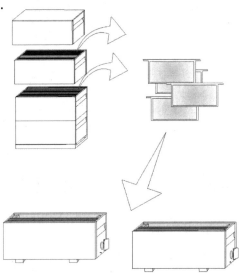

You could even divide the entire hive into nuc-boxes..
(A five box hive could theoretically be broken down into twelve x 5-frame nuclei.)

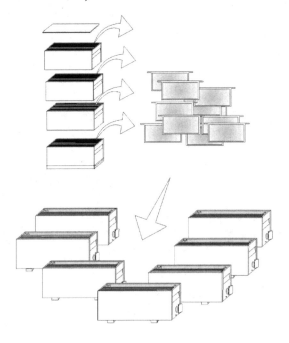

However many parts you divide your hive into, *make sure each has a good mixture of bees, stores and brood in it*. If necessary, swap frames around and shake in bees where appropriate. When you rearrange frames, always leave the ones with brood on them in the middle of the box.

The key to any of these operations is moving all but one of these newly made parts of the original hive onto another site at least two miles away, so the bees re-orientate themselves around their new home and don't just fly back to their old one. After a month or two you can bring all the hives back home again, so you only need to borrow a space for a while.

Rose hives make this part of the job very easy indeed - you'll soon have boxes each with plenty of workers/drones in them. Now they each need queens..

The extraordinary thing is that, provided all these separate parts of the original hive contain at least some eggs or young brood, they would try to make a new queen themselves *and most would succeed*. So you could, if you chose, increase your stocks *simply by dividing up your hives and for the most part you would be successful*. The queenless parts would make emergency queens. How easy is that? (How obliging and adaptable are honeybees?!)

It's true that the queens might not be the best - they would have been made *in* the comb, rather than *under* or *on* the comb where they would have had more room, and they may be made in hives that don't contain enough bees to do a top-class job, but for those beekeepers who want to keep things extremely simple this method is almost foolproof. Anyone can do it, just make sure each piece of the original hive contains young brood and bees, that's all. Be patient, after six weeks, with luck, the new queens should have hatched, been mated and have started laying.

Now, I'm not really recommending this approach (except where people wouldn't do anything else at all) - because we can do so much better and with only a little more effort. We can help these queenless colonies by providing un-hatched princesses for them, giving them a head-start. Not only that, but the princesses will be *planned* ones.

Planned princesses are made by bees who are not in a hurry. They are not queenless, but they choose to make more princesses anyway. These princesses develop in big queen-cells in hives that have plenty of healthy bees and food in them. They are made when the old queen needs replacing (in supercedure cells), and (here's the important bit), also in the run-up to swarming (in swarm-cells).

a hatched swarm-cell

So many people overlook the quality of swarm cells because they've been taught to see them as something unwelcome in the hive, and that's a shame; *swarm-cells are the very best queen-cells bees ever make.* Large, well-shaped cells that contain large, well-fed princesses made in large hives at their prime. What could be better than that?

A lot of nonsense is written about not breeding from colonies that swarm - *every healthy colony swarms!* The best queens anywhere will, given time, eventually get round to swarming. It won't be in the first year, it may not be in the second, but certainly by the third year she will swarm. And in preparation for that her hive will produce perhaps ten or twenty of the best quality queen-cells you could ever hope to find. Some say squash them, I say harvest them and be happy.

(There *is* an argument for not breeding from queens that breed *excessively,* ones that swarm more than once a year, for instance. But if you only breed from your best hives you won't be including any of those queens anyway, because they will never have a chance to build up a colony worth breeding from. My own judgement of which are the best colonies is very simple - they have to be the biggest in the apiary and not too stingy. (Big hives are always more stingy than small ones so it's a matter of proportion.) I know there are lots of other criteria that people assess bees on, but it never bothered me that much whether they run about on the comb, for instance.)

If each divided part of your original big hive is given a capped swarm-cell from one of your best hives (which could be the same one that you're dividing up), then you will have saved the bees a lot of time and trouble - and potentially given them an excellent queen.

So, if your best hive decides it's time to swarm, don't panic, don't despair, instead make the most of this golden opportunity to increase not just your stocks, but your best stock.

I put a frame that has a good swarm-cell on it in the middle of a box and make up a nucleus around it (or fill up the rest of a whole box), shaking in extra bees if needed. If there is more than one really good swarm cell on a frame, I cut the extra ones out with a thin knife and attach them to the top bar of other brood-frames and put each into nuc-boxes. These princesses will hatch within a week and, if all goes well, they could be busy laying eggs two weeks after that.

I also put good swarm-cells into an incubator (or you could use a queenless colony). They need to be in individual cages or the first one out will kill the others. Once hatched they go into poly-nucs (a form of mini-nuc or mating-nuc) where a handful of bees look after them until they're mated and laying. After that I mark them.

a newly-hatched native Irish princess - small and dark and perfectly adapted to my area

These *planned* and *mated* queens can be used to replace failing queens, or as an even better alternative to the swarm-cell in the divided boxes. This is the best option of all because the new hive or nucleus will hardly miss a beat: the new queen will start laying as soon as she's been released from her introduction cage by the bees, and the colony will build up in no time.

There's quite a bit of planning and work involved in getting mated queens ready and I never seem to have enough (although I aim for around sixty a year). It's well worth the effort though, for the head-start it gives the recipient colony.

Just to be clear, I'm not advocating breeding from any old hives just because they're swarming - we should always breed from the best stocks we have - but I am suggesting that if your best hives start making swarm-cells you should make the most of the opportunity.

If a weaker/grumpier hive begins swarm preparation, my policy would be to take the queen out (if she's marked the job is much easier) and, unless she's clearly no good, make up an artificial swarm with her. This is never quite as successful as a natural swarm, but at least it's in your apiary and not in the neighbour's chimney. Just shake her off the frame and onto the floor of a nuc-box, shake in lots and lots of bees, put in some frames of foundation or empty comb, and put the lid on. Don't feed them or leave them any brood - remember, you're replicating swarming. I move them to another apiary, open the door and then forget about them for a month or two, and am often surprised at how well they've done. They may well supercede the queen in the autumn and be ready for full production next year.

As for the rest of the weaker hive: I would divide it up if I had mated queens, or swarm-cells available from a better colony (fortunately, hives in an apiary often all start swarming simultaneously). And if not, I'd reduce the swarm cells inside to one and leave them to it.

I find selective breeding in general, and specifically raising new queens, to be the most fascinating part of beekeeping. I try out different methods every year and I certainly don't just wait around for my best hives to start swarming. Watching bright, inquisitive queens hatching from their cells, watching their mating flight (what I can see of it), finding their first eggs - it's all really interesting and exciting. There are lots of ways of producing queens that might be too confusing and complicated for this book, but the methods out-lined above should get you started and I strongly urge you to have a go. What have you to lose? Yes, you'll make some mistakes, you'll have some abject failures - you have to expect those - but that's why bees are so prolific. They make plenty of everything so there's room for mistakes and failures - and successes, too.

Incidentally, if you're wondering how I check every frame in a hundred hives every nine days from April to August, I don't. Perhaps half my hives have queens born just the year before so, given enough room, they're unlikely to swarm. Of the other half, I only need to check them until I first find swarm-cells. After that, whichever way I've responded, they won't be swarming again that year.

I use a very simple and non-invasive method of checking, too (some might even say lazy), that is, I hinge up each box in turn *very carefully,* peering at the bottom edges of the frames. If I see swarm-cells then, and only then, do I check every frame in the brood-nest. This works well most of the time because Rose frames are reasonably shallow and the swarm-cells are likely to be at the bottom of the frame, or close enough to be seen from the bottom. If I've looked through the top half of the hive (however tall it is) in this way and have found nothing, then I assume there won't be any swarm-cells in the bottom half either, so I stop. I often don't need to remove any frames at all in these inspections and the whole procedure might literally take just a couple of minutes for each hive. (Unless, that is, I actually find swarm-cells..) Of course I miss a swarm-cell from time to time with this method, but that's ok too.

The reason I tip up the box very carefully is because the princess pupae are hanging loosely inside their cells (unlike the pupae of workers and drones) and could fall down and die inside if the cell is handled roughly. If I want to use these swarm-cells, I need to keep them nearly vertical and unshaken at all times.

Chapter Seven

HARVEST

honey is only capped when it's in perfect condition

Honey is not our only beekeeping harvest. Harvests also include the extra nuclei and hives we sell, the candles we might make, pollen and propolis - and also the fruits and seeds we collect from the garden. (Perhaps we could go beyond that to include our health, the countryside around us, the friends we make, a whole new way of looking at the world - but, then again, perhaps that's being a bit too prosaic.) Certainly, working with bees brings all kinds of unexpected benefits to those who make room for them in their lives.

a few of the 100kgs of blackcurrants we harvest every summer

hand-dipped beeswax candles

Honey, though, is still likely to be your most popular harvest - just watch how everyone who smells it smiles..

ready for the shops

Honey is good food and the best is expensive and highly prized. There's always a market for local honey, and if you can write *'My bees have never been fed sugar'* on your labels then you will find plenty of loyal customers. (I would love to see this as a standard requirement for all producers). Though I produce literally tons of honey annually I never have a problem selling it. My customers can taste what they're getting - one of nature's greatest gifts. Around here the summer honey is predominantly made up of blackberry and clover but there are lots of other nectars in there too, like purple-loosestrife, fuchsia and heather, all from wild flowers. What could be better than that?

a Rose box of capped honey weighs around 25kgs (50lbs), yielding around 20kgs (40lbs) of extracted honey. (worth approximately €200 wholesale when it's been bottled in labeled jars)

Harvesting honey can begin whenever there's capped honey - that could be immediately after the oilseed rape if you have it, but for me harvest begins in July with a few boxes off the top of the tallest hives. In late August I begin clearing systematically through the apiaries and the process could go on into October or even later.

Working down from the top, I remove boxes of capped honey, shaking off the bees with a sharp knock-down on the upturned roof (another reason why roofs need to be strong). This only works if you first take out one frame and shuffle the others along to make a little room between each one.

full-size hives like this one can yield 100kgs of surplus *honey*

Sooner or later, and you could take off 50 - 100 kgs of honey before you reach this point, you will come across either uncapped honey, or the top edge of the brood nest, or both. That's where to stop. Remember, you can't extract un-capped honey because it isn't honey yet and it will just go mouldy or start fermenting in the jar.

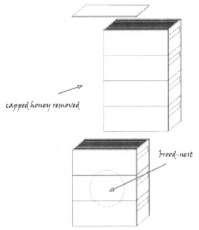

If you reach a box that is mostly capped honey, with just, perhaps, the middle three or four frames containing brood, then use some honey-filled frames from the box below to make up the full box. In this way you could exchange frames until you had a whole box of capped honey and everything else was left on the hive. Remember to leave any frames that contain brood in the centre of the box.

I find my flat-bed barrow invaluable

This process of separating and sorting frames as you go down through the hive is far simpler than it sounds. You certainly don't need a queen-excluder, you just need to have a quick look at the combs as you go; white and crinkly means honey you can take - *anything else* means something which should be left behind. *If in doubt, leave it behind.* Remember, you need to leave more honey for the bees than they can possibly eat.. *(See 'Winter Preparation'.)*

this photo shows the unmistakable difference between capped honey (left) and capped brood (right)

If your hive is still three or four boxes high when you've taken all the capped honey off, it may be worth coming back in a couple of weeks to see if there's another box ready.

18-frame radial extractor (with covers removed)

Rose frames fit into any extractor except the very smallest ones designed specifically to take four National super frames. Any extractor bigger than that (and the vast majority are) will take at least some Rose frames. After the frames have been through the extractor you'll get a chance to inspect the combs, sorting the good from the bad. This time, if in doubt, throw it out.

(I am asked occasionally whether harvesting honey from combs that once held brood is healthy and legal. The answer to both questions is *Yes*. Because it's never possible to extract the last drop of honey from a cell, there's always a lining of honey left behind, and this presumably seals in any contaminants there might be in the cocoons. Even in a hive with an excluder there is a strong likelihood that the honey has spent at least some time in a brood-cell anyway, before being moved up the hive.)

Most sorts of honey will crystallize, wherever it's left for any length of time; in the comb, in the settling-tank, in the storage buckets, or in the jars. At this point it will have to be warmed up to liquefy it again in order to move it or strain it. Warming, though, however carefully it's done, has an impact on the flavour and the fragrance of the honey, as well as on the nutritional value, so ideally this should be avoided.

To get around this I try to get all my bottling done as the honey's extracted, producing 'cold-filtered' honey. So as each apiary is harvested I try to process all the honey into jars before I start on the next one. The problem is that cold honey that's close to crystallizing, although absolutely delicious, is sluggish and reluctant to go through fine-mesh filters. So I don't bother trying. I settle it overnight and then run it through open-mesh strainers and accept that small bits and pieces go through, too. I warn customers with a little label on the lid explaining that they may find wax and pollen - perhaps I have particularly tolerant customers but no one seems to mind.

I'm telling you this to encourage you to at least try selling 'raw', cold-filtered honey - offering your customers honey in it's healthiest, tastiest form. It will go on to crystallize in the jar but then at least it stays on the toast better.

There are plenty of people who will only buy runny honey, so I do have to warm some honey up for them, but it's not my preferred way.

If you have just a few kilograms of honey to harvest then you won't need an extractor at all. Just eat it along with the comb. Or else scrape the combs down to the wax foundation sheet with a hive-tool and put the honey and broken comb into a clean muslin bag (or a jelly bag, or even an old pillow-case), and leave it to drain overnight in a warm kitchen.

Or you could do what people have been doing for millions of years - just stick your finger in!

provided temperatures get high enough, white clover is an excellent honey plant

Chapter Eight

WINTER PREPARATION

Preparing your hives for the winter amounts to making sure they have plenty of honey but very few Varroa mites.

Stores

I'm often asked how much honey to leave for the winter and the short answer is *lots*. Leave more honey than your bees can possibly need, just in case, and none of it will be wasted. If the bees haven't eaten it by next spring, then you can harvest it then.

I never feed my bees sugar. That's because bees don't eat sugar - they eat honey. They are not the same thing.

Honey contains:	Sugar syrup contains:
Fructose, Glucose, Sucrose, Maltose (and other sugars), Proteins, Vitamins B2, B3, B5 B6, B9, C, Calcium, Iron, Magnesium, Phosphorus, Potassium, Sodium, Zinc, Sulphur, Manganese, Silicone, Copper, Chromium. And hundreds of other things like amino acids, enzymes, hormones, acetic acid, citric acid, volatile oils, flavours and tastes.	Sucrose

Traditionally, beekeepers took all the honey they could and then replaced it with sugar syrup. This is not just greedy it is also short-sighted and silly because bees need all the complex foods in honey to stay healthy. Feeding sugar also gives beekeeping a bad name because customers are (rightly) worried that perhaps what they're buying in their honey jars isn't all honey. If you feed sugar, even in the autumn, how can you possibly guarantee that none ever gets into the honey harvest? And yet this approach is still taught almost universally.

Every single year, since bees first were bees, they've collected and stored enough honey to get them through the winter. They know what they're doing. They don't need sugar. Leave them all the honey they need, and more, just in case, and they will emerge next spring strong and healthy.

(Of course, if your colony is close to starvation, for whatever reason, then obviously they'll need feeding, but this should be an extreme exception. No one should feel bad about feeding their bees if they need it, or hesitate for a moment, but to feed routinely, without thinking, seems to me to be a mistake. And if bees need extra feed because too much honey's been taken off, well then someone's got their priorities back-to-front.)

Some people will argue that sugar does no harm at all to bees, and we might discuss the subject at length, but perhaps the real discussion ought to be - What's the best possible food for our bees? How can we ensure our bees are getting all that they need in their diet? Surely there can be only one answer: their own honey, collected from a wide variety of local flowers and stored in a clean hive.

Very fortunately, in almost every year bees make and store *extra* honey, way beyond what they could possibly need - that's the bit we can take. No more than that.

(Only once in all my years of beekeeping did I not get a honey harvest. That was in 2009. It started raining here in early June and didn't stop till mid August. It was the wettest year here since records began in 1830. Incredibly, there was a little honey but I left it for the bees. The winter that followed was the coldest and longest since 1947. Of course I worried because, as always, I hadn't fed any of my hives. I lost only one small nucleus to starvation out of around 100 hives.)

In practice, I would over-winter a big colony in three or four Rose boxes, most colonies in two, and plenty in just one box. In each case the boxes would be perhaps half full of capped honey at harvest time but stuffed to the gills by October because of the autumn flowers, especially ivy.

Often the bees will have eaten only half of the stores over winter, but sometimes more. I harvest whatever's in the way in April - they will be getting fresh stores in by then - and sell it as 'Autumn Honey'. It is a darker, stronger honey - perhaps because of the ivy? - and some people love it.

I can hear people saying "It's alright for Tim! His bees have wild flowers to forage on through the autumn. We have nothing around here!" The truth is that every area has advantages and disadvantages. Here the bees have to cope with frequent Atlantic gales and annual rainfall of over two metres, they can't forage over large areas because it's under the sea and there's no commercial agricultural or horticulture beyond forestry and grassland. On the other hand, there are small fields still, with glorious overgrown hedges filled with flowers of all sorts, so if the weather's kind they can find something between February (willows) and November (ivy).

Perhaps you have oil-seed rape, or field beans, or soft-fruit orchards or heather moors; perhaps, if you're farther south, you have acacia woods, or sunflowers; perhaps you have none of these but instead a rich suburban landscape that yields something all year round. Whatever your circumstances, there will be different harvests for the bees throughout the year. Whenever the last one ends, the bees will need to have stored enough honey to last till the next one begins again. If your last harvest occurs in August, and there's nothing after that until the following May (highly unlikely), then you'll just have to leave them enough in August to last them right through. I would guess though, that if you watch closely, you'll always find something that bridges that gap.

Varroa

If you haven't treated against Varroa by the autumn you will probably need to. If you have looked *very* carefully and are convinced your bees are free of mites, then leave them alone, otherwise choose one of the many treatments available and get the job done. (The Beebase website describes ways of monitoring your Varroa populations - *see 'Resources'*.)

After the honey has been harvested you're free to use just about any of them (except those that harm brood - for those, like gaseous Oxalic Acid, you'll need to wait till mid-winter), but always follow the instructions carefully. You may be tempted to try home-made preparations, and there's nothing wrong with that at all, *provided they actually work,* and provided they don't harm your bees (quite a challenge..).

Even if you are a completely minimalist, hands-off beekeeper, and haven't done any of the other beekeeping tasks throughout the whole year, protecting your bees against Varroa mites is the one job you really need to do. Otherwise your bees will die.

Mouse Guards.
Don't forget the mouse guards!

a family of mice living in a box of combs

Winter

After that, perhaps put a stone on the roof, wish them well *and then leave them alone until next April.*

Appendix One

CONSTRUCTION PLANS

You can buy Rose hives easily now - but you might like to try making them.

I developed this construction method years ago and it's so simple I don't know why all hives aren't made like this. Unlike other methods which seem to use a dozen or more pieces for each box, all you need for this one are four pieces. And two of those are just simple rectangles of plywood. You'll need a reasonably equipped workshop, but you really can make these hives yourself..

The Box
A Rose box is made up of two pieces of 12mm ply and two pieces of profiled timber..

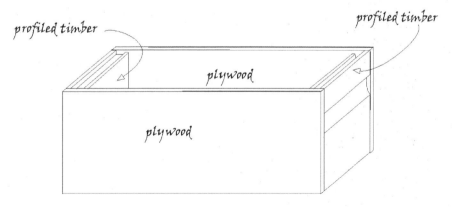

Each piece of ply measures 460 x 190 mm
Marine-grade ply is probably best, but WBP works well too.

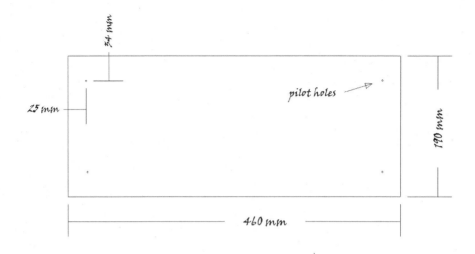

Make sure your diagonals are equal otherwise you won't have a true rectangle.

Four 2 mm pilot holes should be drilled, as indicated, near the corners..

The two profiled-timber pieces can be cut from any suitable soft-wood - Cedar is probably best, but Douglas Fir, and good-quality white deal (Spruce), which is cheap and light-weight, can also be used.

I use the best 9x2" white-deal floor-joists I can find at the local builders' suppliers, picking the ones with the tightest grain and cutting them to length.

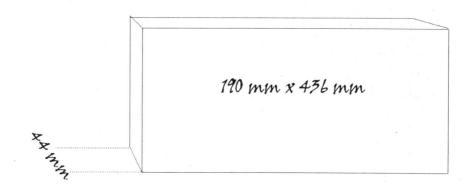

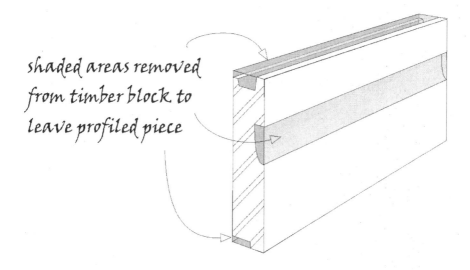

shaded areas removed from timber block to leave profiled piece

　　　　The rebates which turn the simple block of wood into the profiled piece should be cut along the whole length of timber - right to the ends. (A spindle moulder makes the job really easy, but see below for a simpler version if you don't have one of those.)

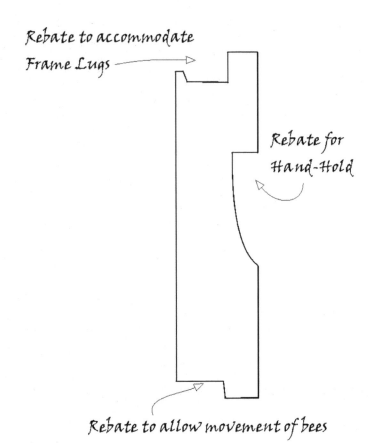

This might look complicated at first glance, but just take one rebate at a time..

It's important to get the top rebate right - you need to allow room for the frame to sit down inside the box comfortably, but not too deeply or it will reduce the bee-space below it.

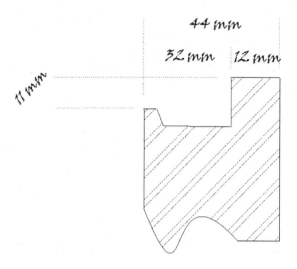

Follow these dimensions as accurately as you can..

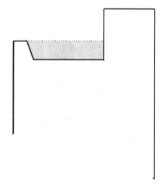

The space marked on this drawing is the small gap under the frame lugs - you'll squash quite a few bees without it when you replace frames. It only needs to be 5 mm deep..

The other two rebates are not as critical - the one at the bottom should be deep enough for bees to move around in, and the hand-hold should be comfortable for the user, remembering that your box will soon be full of honey and weigh a lot.

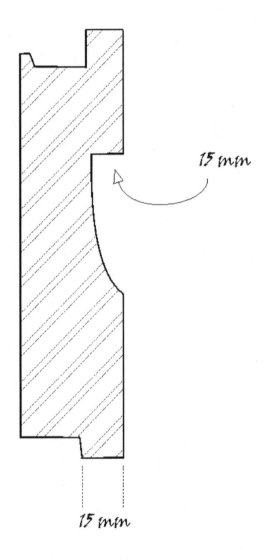

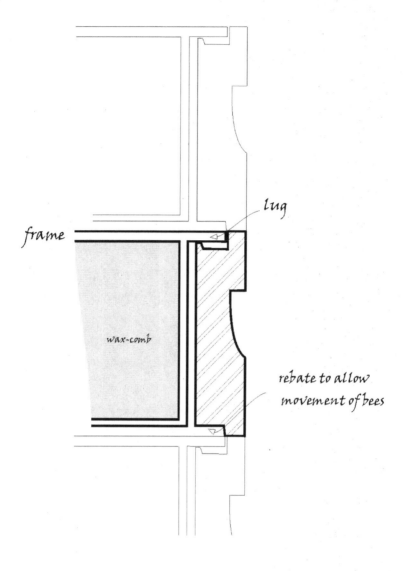

This is the way the frames will sit in a stack of boxes - they're arranged so there's a gap for the bees to move freely above, below and all around each one..

ALTERNATIVE PROFILE
for those without access to a spindle-moulder

Both of the rebates can be cut with a table-saw, router, or shoulder-plane.

A plastic frame runner (from your supplier) is simply nailed on.

Make sure the gap left above the frame runner is big enough for your frame-lugs - 10 or 11 mm.

The hand-hold rebate is replaced with a full-length batten. This is much easier to make but it adds a little to the weight instead of reducing it.

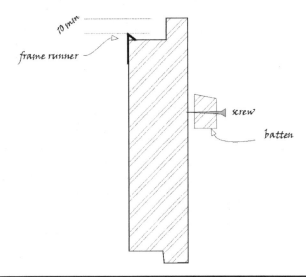

Now you're ready to put the box together. All you need is eight x 40 mm woodscrews and some water-proof wood glue..

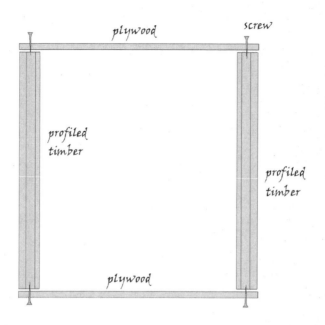

the four parts of a box ready for assembly

Obviously this is a very simple construction method, and one that's very strong, - but please note the following
IMPORTANT POINT:

A Rose box (like any hive box) has to be made not just square, but level and un-twisted too. Otherwise, there will be a gap between boxes in a stack...

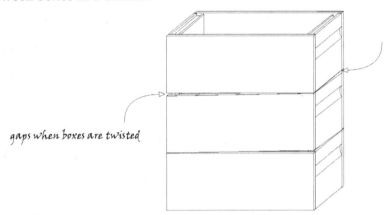

gaps when boxes are twisted

I recommend the following assembly sequence to ensure there's no twist..

Step 1. Apply glue to both ends of a profiled piece..

Step 2. Attach the ply pieces *but use only the top screw in each piece*

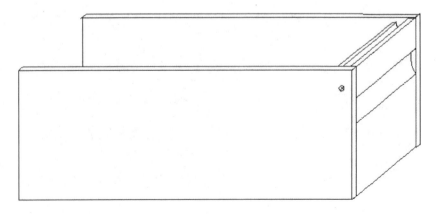

Step 3. Apply glue to the other profiled piece and fit it between the ply pieces, *again only using the top screws..*

Step 4. Stand the box on a firm surface, squat down until your eyes are level with the top edges. Check that the far edge is parallel with the near edge.

If not, tap down the lowest corner onto the table/bench until it is..

This little step really is important and hardly takes a moment..

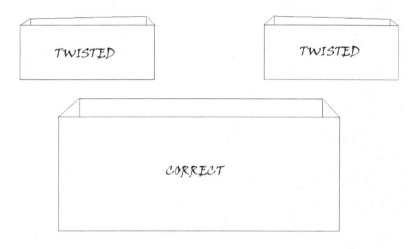

Step 5. Put in the last 4 screws.

That's it - you're done. When the glue is dry, treat and seal or paint your box to preserve it. Remember, though, that many preservatives are toxic to bees..

LEGAL NOTE:

Both the Rose hives and this construction method (whether used for Rose boxes or any other hive boxes) are patented (S2008/0730, and S2008/0731). Their commercial use and distribution are strictly prohibited except by license.
However, full permission to build as many Rose hives as you like is hereby granted, provided they are for your own use only.

(Get in touch if you're interested in a license for commercial use as manufacturer or supplier outside the U.K. and Ireland)

Your Rose boxes will fit Commercial or National roofs, crownboards, excluders and floors because they're all 460 mm x 460 mm square. Plans for those are readily available and they're not difficult to make. If you have succeeded in making a Rose box then you will certainly be able to make the other parts you need.

The Frames

The frames have the same top and bottom bars as a standard National frame, i.e. they're 432 mm along the top and 356 mm along the bottom.

The side bars are 182 mm long.

Frames, though are quite complicated to make and perhaps only worth trying if you're really up for a challenge! Rose frames are available from Thornes, as is Rose size wired foundation.

Remember, you'll need 12 frames for each box..

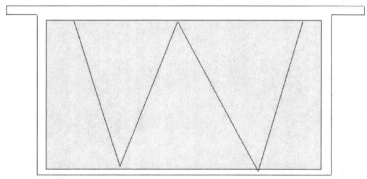

frame with full sheet of wired foundation

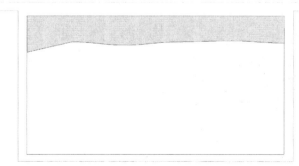

frame with strip of un-wired foundation for 'free comb' building

Nucleus Boxes

If you can make a Rose box from scratch, or put one together from a kit, then you can certainly make one of these nucleus boxes, too. The 5-frame box is strong enough to transport bees, collect swarms and even over-winter nuclei. I use them often and have about 40 of them. I'd certainly recommend you keep a couple spare at all times, ready for that beautiful natural swarm, or that timely artificial swarm, or to make up a nuc around a perfect queen-cell. I never leave the house without at least three or four in the van.

5-frame nuc-box

They are made using the same ply-wood side pieces as the whole box, but have shorter end-pieces. Just cut one of the standard-length profiled-timber pieces down to give you two lengths each 176 mm. *(If you make your own profiled pieces, leave out the bottom rebate.)*

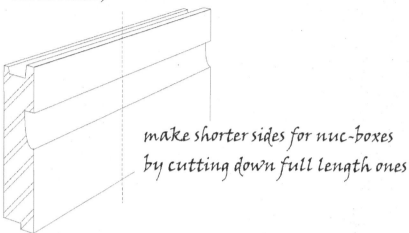

make shorter sides for nuc-boxes by cutting down full length ones

They need a crown-board too, obviously, and a loose-fitting roof.

The floor is best made from a cut-down floor-mesh. Put a couple of battens across the bottom to keep it up off the floor, otherwise there will be no ventilation.

The door way is just a large hole drilled at one end. The door is a piece of ply fixed loosely at one corner with a screw, so it can be flopped over in front of the hole. A piece of foam sponge works well too.

If you cut the end pieces to different lengths you'll get boxes which can accommodate any number of frames, e.g. -

1 frame - 40 mm
2 frames - 74 mm
3 frames - 108 mm
4 frames - 142 mm
5 frames - 176 mm
6 frames - 210 mm

Appendix Two

CONVERTING FROM OTHER HIVES

It's tricky and awkward changing hive-types - that's why people don't usually do it, and end up stuck with the hive they happened to start with many years ago. It is perfectly possible, though, and basically involves only adding Rose boxes when the colony's expanding and removing as many non-Rose boxes at the end of the season. Here are a couple of suggestions for going about this..

For replacing brood-filled frames without harming the brood, gradually work them up the hive until the brood inside hatches and is replaced with honey.. Then take them off, extract the honey and melt down the wax. (You'll know by the weight of a National brood-box full of honey why I arrived at a smaller box!)

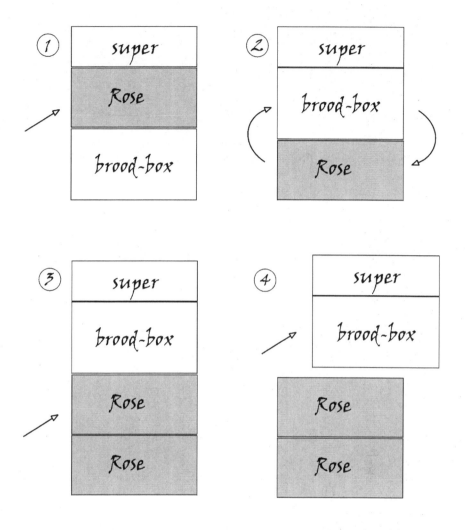

Don't rush this process; it could take a season, or longer. Make sure the Rose box is nearly filled before you take the next step.

If your National supers are full of good comb after extraction and you really don't want to waste it, put the frames into Rose boxes, alternating them between Rose frames, like this..

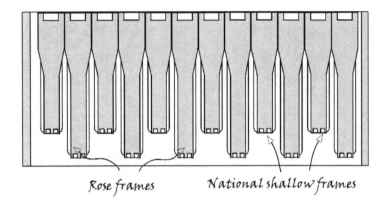
Rose frames   National shallow frames

The bees will extend the combs in the shallow frames until they fill the void, and you can count them as Rose combs until they become too old..

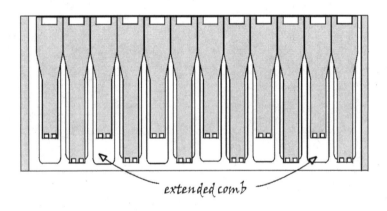
extended comb

Unless they're at the bottom of the stack, though, the extra comb that's built could also be stuck to the frames below, so this method isn't ideal, but you could give it a try and see for yourself. (You could put super frames into a Rose box before they're extracted too, obviously, but don't do it with frames with brood in them, because it might get chilled.)

National deep frames that have good combs in them don't need to be wasted either - after extraction, run them carefully through a table saw set at 182 mm. The bottom-bars will be cut off as well as some of the comb, but what remains should be strong enough to support itself if handled reasonably. (This probably should not to be tried with wired combs.)

Both National deep and shallow frames can be taken apart and rebuilt as Rose frames by using Rose side-bars.

Some people have told me they cut out combs from other frame-sizes and put them into Rose frames. A good fit helps hold them in place till the bees fix them properly, or you could use a couple of rubber bands. This is messy and disruptive, but I'm sure it would work and you'd get the transfer over and done in no time.

Appendix Three

STARTING FROM SCRATCH

If you're not a beekeeper yet, and only planning your first hives, then I'm guessing that either you're a very fast learner, or else you're totally confused by now. Either way, well done for persevering with this book!

As I'm sure you've noticed already, one of the biggest challenges for aspiring beekeepers is deciding who to listen to. There are so many books out there and all have a different take on the subject, and you will be bombarded with advice from all quarters. Actually, you'll quickly realize that a lot of the books you'll come across are simply re-hashed from earlier ones, with a fancy new cover but no new ideas. There's quite a bit to learn and I'm not going to attempt to teach you anything here, only encourage you to get a good book and make some beekeeping friends.

Early on, though, you'll have to decide which hive-type to go with - not an easy decision when you can't possibly know all the implications. Please don't just go with Rose hives because I say they're the best - get a second opinion at least. Ask someone who *really* knows bees to look at the ideas and methods here, or ask at your local beekeepers' association. It's certainly not the end of the world if you make the wrong decision but changing hives later is complicated and you should be happy with your choice from the beginning.

If, after that, you choose Rose hives, well then, good for you - and good for your bees. I shall do my best to support you where I can through the website *(see 'Resources')*, where you'll be able to talk to others in the same position.

You'll need to buy or make a hive or two - and then get some bees. That could be the tricky bit. 90% of the people who come to my courses don't have bees yet, and apparently most members of beekeeping associations don't have bees either - a sure sign of the extraordinary and worrying times we live in. You may be lucky and live near enough to another Rose hive user, and be able to buy bees from them already installed on Rose frames, but if not, you'll need a different approach. (Frames from other hive-types wont fit into a Rose hive.) Here are your best options:

1. Collect a swarm (or, more accurately, a cluster). Make sure you have a suitable bee-proof, *ventilated* collecting box to hand at all times (a 5-frame nuc box is perfect - *see 'Plans'*) and then tell everyone you know that you're willing to collect swarms. Leave your phone number with your local council, the fire-brigade, the guards (police); put notices in the paper and local shops - and then stand by your phone. You will get lots of false alarms - wasps' nests, bumble-bees, honey-bee colonies that have already moved into a new home and have started building combs (leave those alone until you've had a bit more experience), but sooner or later you will come across a cluster which you can take home for free.

2. Explain to your beekeeper friend that you don't just want bees, *you want bees on Rose frames, please.* (You might have to show him/her this book.) Persuade them to let you leave a Rose box in the middle of one of their hives for three or four weeks in the summer (they shouldn't have any objections to a clean new box) and then collect it when the bees have expanded into it. (Your friend will deserve a generous thank you, of course.) Obviously your box will need to be placed below the queen excluder so there will be plenty of brood in it. When you collect the box (you'll need

to strap it up with a floor and crown-board) and take it home, they'll probably be queenless so you'll have to introduce a new queen or let them make their own one.

3. A shook swarm. Again, you'll need the co-operation of your beekeeper friend. They'll need to shake lots of bees into your Rose box (or nuc-box) and add the queen. This works best when they're about to swarm. You might like to explain to them that they could easily lose the swarm anyway and this way they'll at least gain a couple of bottles of wine. Take them home and feed them if you like but leave them shut in for a couple of days before releasing them.

4. If all else fails you could always buy a colony of bees in either a Commercial or National brood-box and then only use Rose boxes for expansion. Following the hive management methods from earlier on, you'll eventually work the brood-box out of your hive and the job is complete.

Having only one hive of bees is a precarious situation to be in, so I'd suggest your priority should be to divide it in two when the colony's big enough. But once you have bees in a hive then you're a beekeeper. Congratulations - the rest is easy!

Appendix Four

RESOURCES

ROSE HIVES WEBSITE:

www.rosebeehives.com
Here you'll find more photos, a forum, etc. This is also where I list any
beekeeping courses and colonies on Rose frames.
Please note - I don't sell hives except with bees (just too busy..!)

SUPPLIERS:

(Britain)
E.H. Thorne (Beehives) Ltd - www.thorne.co.uk
(and their agents)

(Ireland)
Gearoid MacEoin - www.maceoinhoney.com
(he'll post out what you need)

(If you live outside Britain and Ireland and you'd like to be a manufacturer or supplier of these hives, do get in touch.)

YOUR LOCAL BEEKEEPERS' ASSOCIATION

(Ireland)
www.irishbeekeeping.ie

(Britain)
www.britishbee.org.uk

BEE DISEASES TESTING

Pat Maloney
Teagasc,
Kinsealy Research Centre,
Malahide Rd,
Dublin 17
(send a sample of bees/suspect comb along with €5)

In the U.K. this is done through local bee inspectors who will find you once you've registered with BeeBase
www.secure.fera.defra.gov.uk/beebase

If you register as a beekeeper in Ireland through the Dept. of Agriculture www.agriculture.gov.ie you will be sent a copy of Dr Mary Coffey's book 'Parasites of the Honeybee'.

In the U.S. each state has its own agency.

BOOK RECOMMENDATIONS

GUIDE TO BEES AND HONEY, Ted Hooper, Northern Bee Books
BEES AND HONEY, Michael Weiler, Floris Books
A WORLD WITHOUT BEES, Benjamin and McCallum, Guardian Books
THE BAREFOOT BEEKEEPER, Phil Chandler, www.biobees.com

P.s. I've tried to explain in this book how urgently I see the need for changes in beekeeping, and at the same time how we can easily make those changes. If you have any questions or comments to make then do, please, get in touch. Consider this as a sort of village hall presentation which will be followed by an open discussion, just as if I'd been standing in front of a projector waving my arms about and rabbiting on, as I often do. (Where are the tea and sandwiches?) It's your turn now to have your say if you like - on the Rose Hives website forum - www.rosebeehives.com.

If you disagree with the ideas here then explain to everyone why - that's how we'll all learn. If, on the other hand, you think this book could be useful to other beekeepers in their quest to be better at their job, then please pass it on. Otherwise no one will hear about it and the standard beekeeping hives and practices will continue, along with the miserable decline in bees. You might even feel inspired to write a review on a blog, in the newspaper, or on Amazon.. (And, hey, if you write a book on bees or beekeeping I promise I'll buy it and read it, too..!)

P.p.s. Along with my ambition to keep bees in the best way I can, and to raise extra stocks every year to spread around, I also consider it my responsibility and pleasure to share what I've learned about bees over these last few decades. While this book is for people who already know what they're doing, lots of the people I meet are just beginning their journey to becoming confident beekeepers, so these days I teach people the basics of beekeeping in various different courses. It's hard work, great fun - and also very satisfying to see people getting excited as they begin to understand both the intricacies and the vastnesses of the subject.

Anyway, I'm happy to report (in the interests of beekeeping as well as my finances) that these courses are usually over-booked, such is the demand. So I'm hoping to bring out a companion title soon called The ROSE HIVE Method - For Beginners based on what I teach and what I've learned on those days...
Tim

guess what we're doing!